1
2

Charles Darwin: Origins and Arguments

www.pocketessentials.com

Other titles by Bill Price

Celtic Myths
Tutankhamun

Charles Darwin: Origins and Arguments

BILL PRICE

POCKET ESSENTIALS

First published in 2008 by Pocket Essentials
PO Box 394, Harpenden, Herts, AL5 1XJ
www.pocketessentials.com

Edited: Nick Rennison
Indexed: Richard Howard

A CIP catalogue record for this book is available from the British Library.

ISBN 978-1-84243-312-6

2 4 6 8 10 9 7 5 3 1

Typeset by Avocet Typeset, Chilton, Aylesbury, Bucks
Printed and bound in Great Britain by J.H.Haynes Ltd, Yeovil, Somerset

Contents

Introduction **7**

1: The Book That Changed the World **13**

A Victorian Gentleman; Darwin's Big Year; On the Origin; The Controversy Begins; The Philosophical Naturalist

2: The Making of a Naturalist **55**

Early Years; Edinburgh and Cambridge; A Five-Year Mission; A Twenty-Year Wait

3: Evolution After the *Origin* **107**

Darwin's Later Years; Darwin and Society; Evolution Today

4: The Controversy Continues **131**

Evolution and Christianity; Monkey Trials; A Grandeur in this View

Notes **147**

Books by Charles Darwin 153

Bibliography 155

Index 157

Introduction:

One of the characteristics of humanity throughout our history has been an almost insatiable need to question ourselves in an attempt to find answers to the unknowable aspects of our lives. We appear to require meaning, to know where we are from and what we are doing here. The religions of the world have dealt with these traits of human nature by providing systems of understanding which are based on faith and belief. Since the beginnings of modern science during the Renaissance, when the knowledge of the Greeks was rediscovered, this has led to a conflict of ideas between theology and rationality. This is nowhere more apparent than in the continuing debate over what is now generally called Darwinian evolution, after the man who has become the figurehead of a revolutionary change in human thought.

As revolutionaries go, Charles Darwin cuts an unlikely figure: a Victorian country gentleman of independent means, an amateur naturalist, a devoted husband and father. He is not often compared to his near contemporary Karl Marx, but the impact of the publication of *The Origin of Species* was nothing short of revolutionary in the influence it would go on to have on the biological sciences and, more generally, on how we as human beings perceive ourselves and our place in the world.

In the history of thought, few theories have had such a fundamental effect on humanity or been so controversial. Darwinian evolution provoked a furious response from its critics when the *Origin* was first published in 1859 and now, 150 years later, the debate continues, and the positions on both sides of the argument seem as intractable and entrenched as they have ever been. Perhaps it is a sign of the importance of the issues at stake, which go to the heart of what it means to be human, that what is essentially the same argument has gone on for so long, particularly as the evidence in favour of evolution is now so overwhelming.

At the heart of Darwin's work was the theory of variation by natural selection, which he outlined in the *Origin*. The theory itself is really quite straightforward and can be expressed simply in a few lines. Recalling the first time he read about natural selection, Thomas Huxley, Darwin's great public defender, remembered being astonished at how obvious it was once he had read about it and how stupid he had been for not thinking of it himself. The idea of evolution – the notion that species of animals and plants changed over time – was not, in fact, new at all during Darwin's day. It had been put forward in the eighteenth century by, among others, Buffon and Lamarck, the leading French naturalists of their day. Darwin's own grandfather, Erasmus Darwin, had also written on the subject. The difference between previous ideas about evolution and Darwin's theory was that, in natural selection, Darwin provided a mechanism by which change in species could occur entirely by natural means. This is what sets Darwin's work apart from the speculative theories of his predecessors and why it has come to be so crucially important.

This may seem something of an overstatement of the impact of a theory which describes how the variety of life on our planet has arisen, but there can be no doubt of its continuing relevance. At a time when we are having an unprecedented effect on our environment, Darwin's ideas that all of the natural world belongs to what he called a 'tree of life' – with each branch being connected to another and humanity being an integral part of the whole rather than separate and above it – are as important now as they have ever been. The Earth, Darwin showed us, is not simply there to be exploited for our own gain, but is at the centre of a system on which all life, including our own, depends.

As both the bicentenary of Darwin's birth and the 150th anniversary of the publication of *The Origin of Species* approaches, it is, perhaps, as good a time as any to look back at Darwin's life and work and to consider its continuing relevance. But this book is not intended to be a straightforward biography of the man or a book about the science of evolution, although it contains relevant details of both. My aim is to trace the development of Darwin's theory and to place it within the context of the period in which he was working.

As a starting point, I have chosen to begin in 1859, with what I have called 'Darwin's Big Year', the year that he turned fifty and that the *Origin,* his major work and the book that changed the world, was published. From there, the book goes back to his early life to examine those important events leading him towards natural selection, particularly the five years he spent on board HMS *Beagle*, an experience which had a profound influence on him and which laid the foundations of the work he would continue to do for the rest of his life.

The book goes on to look at evolution after Darwin, how it moved forward and was, in some cases, misappropriated. It then considers why it remains so controversial today. Perhaps it is a symptom of the fractured times we live in that those who refuse to accept Darwinian evolution use a version of the argument from design in an attempt to discredit it. This argument – that the complexity of the natural world could only have arisen through the actions of a designer, or God by any other name – was current more than two hundred years ago. It has been shown to be unsustainable any number of times, including by Darwin himself, but it continues to resurface again and again. The main area of conflict has been the American courts where legal battles have raged over whether creationism can be taught alongside evolution in the classroom, going back to the so-called 'Scopes Monkey Trial' in 1925 and continuing today in Louisiana, which has recently passed a law to allow the use of creationist textbooks in its schools.

In recent years a counter offensive has developed among the opponents of Christian fundamentalism, who see religion as having a dangerous and damaging influence on society. The evolutionary biologist Richard Dawkins has been at the forefront of this counter offensive and, although the argument has been about more than just evolution, the theories articulated by Darwin 150 years ago remain at its core. It is for this reason, and because of the need to increase our understanding and appreciation of the world in which we live, that this book examines the origins of evolution and the arguments which continue to surround it.

The Book That Changed the World

A Victorian Gentleman

By 1859 Charles Darwin was one of the best known naturalists and geologists in Britain. Since returning from his five-year voyage around the world on board HMS *Beagle* in 1836, he had written a number of highly regarded books, including works on coral reefs and barnacles which remain relevant today, and numerous academic papers on a variety of biological and geological subjects. His account of the voyage, the *Journal of Researches* which is now usually called simply *The Voyage of the Beagle*, had also brought him to the attention of the general reading public, as it went through numerous editions and became what we would now call a bestseller.

By any standards Darwin was a prolific writer. During his lifetime he wrote more than six million words. In addition to the published work, he kept extensive notebooks and journals and also maintained a huge correspondence, keeping in touch with many of the eminent men in his field, including Charles Lyell and Joseph Hooker. He also wrote numerous letters to a diverse group of people – including livestock breeders, bee keepers, pigeon fanciers and gardeners – requesting information he felt could be useful for his work. There are something like 14,500 letters in existence, and probably many thousands more which have not survived, and the extent of this correspondence suggests that Darwin was anything but a reclusive and retiring man, as he is sometimes portrayed.[1]

In fact, Darwin gives the impression of being both sociable and genial. As well as having a large and extended family, he maintained a wide circle of friends, many of whom he met regularly both at Down House, the house near Bromley in Kent which he bought with his wife a few years after they were married in 1838, and during trips into London. Down House, now owned by English Heritage and open to the public during the summer, is in a relatively quiet country location, while being no more than fifteen miles from central London and only a few miles from the nearest train station.

The impression of him as someone withdrawn from society most likely comes from his reluctance to attend public meetings and society dinners, both of which he professed to hate. Instead he preferred to stay out of the limelight at home with his family and get on with his work. As anyone who has ever done an extended period of research and writing knows, constant interruptions and distractions can be extremely irritating. The best conditions for writing are peace and quiet and this was exactly what Darwin created for himself at Down House.

Darwin often excused himself from public engagements because of a recurring illness that dogged him for much of his adult life. It is tempting to think this was simply his way of avoiding a function he didn't want to attend but, at the same time, there is little doubt that he suffered frequently from ill health. The exact nature of the illness has never been fully diagnosed, despite considerable speculation on the subject. In his letters and diaries Darwin described numerous different symptoms, including stomach pains, vomiting, faintness and fatigue, leading to conjectures that he had picked up a disease while travelling in South America, possibly after being bitten by an insect. Another line of thought suggests that his health problems were largely psychosomatic or stress-related responses to overwork. It is also possible that he was something of a hypochondriac who enjoyed the attention he received when he was ill.

Whatever the truth of the matter, Darwin preferred the home life of a Victorian gentleman to the bustle of the city and he did his best to avoid the academic world. Both his family and his wife's family were wealthy, and therefore he was able to follow his own path without ever actually holding a position with an academic institution or, in fact, ever having any sort of job at all. When the Darwins were first married they had a combined private income in the region of £1,300 a year, derived from investments made by both their families. This was a considerable sum for the period and enabled them to pay £2,200 for Down House, where they employed a butler, a nurse for the children and a number of maids and gardeners.

The marriage between Charles and Emma was the third to be made between two illustrious families, the Darwins and the Wedgwoods. Darwin's father Robert and his older sister Caroline were both married to Wedgwoods and

Charles and Emma, who had known each other since childhood, were cousins, and were both grandchildren of Josiah Wedgwood (1730–1795), who founded the famous Staffordshire pottery that bears his name.

The close relationship between the two families was the cause of some concern to Darwin, involved as he was with the study of inheritance. Three of Charles and Emma's children died young – two before their second birthdays and another, Annie, who Darwin would later say had been his favourite, when she was ten. The fact that none of these three died from what would now be described as a genetic disease, passed on to them by their parents, taken together with the fact that Darwin's seven surviving children showed no sign of any affliction, would suggest that the family were in relatively good health for the period in which they lived. At the time child mortality remained high, with medical science yet to get to grips with a number of infectious diseases then prevalent in the country.

A number of Darwin's biographers have suggested that Charles and Emma's marriage was, in effect, a marriage of convenience between two connected families. Charles, after returning from the *Beagle* voyage, decided it was time he got married and selected Emma because she fitted the profile of the type of woman he was looking for. It is impossible to say now if this was actually the case but, if it was, then the couple appear none the less to have had a long and, for the most part, happy marriage. If they were not head over heels in love when they married, then they were certainly devoted to each other and to their children. Unlike Charles, Emma remained devoutly religious throughout her life and it is hard to believe that Charles's work did not cause a certain amount of tension between them. But, despite the deaths of their three children and Charles's frequent bouts of ill health, they appear to have been an essentially happy couple.

Darwin's Big Year

The title of this section refers to 1859, the year *The Origin of Species* was published, but the actual story behind the book does not fit quite so neatly into the parameters of one year. The events that would culminate in the publication of the *Origin* in November 1859, making that year what we would now consider the most important year of Darwin's life, actually began in the middle of the previous year.

June 1858 was a traumatic month for Darwin, both personally and professionally. In the middle of the month his daughter Henrietta caught diphtheria, a contagious disease of the respiratory tract which, at that time, resulted in the deaths of around ten percent of the people who contracted it. Henrietta would gradually recover, but much worse was to come. Almost immediately after Henrietta's illness Darwin's youngest son, who was eighteen months old, fell victim to the scarlet fever epidemic which was then sweeping through the south of England, and died on 28 June.

While the family were enduring this desperately trying situation, Darwin received a letter from Alfred Russel

Wallace, an occasional correspondent who was on a specimen-collecting expedition in the Malay Archipelago. The letter contained an essay detailing Wallace's theories on species change, which bore striking similarities to Darwin's own work. Darwin had been doing extensive research on the subject since he had first articulated his great idea in his notebooks twenty years previously, but he had not made any of his findings public. He had only discussed them with a number of close friends.

Wallace was fifteen years younger than Darwin and did not have any of the social or financial advantages enjoyed by the older man. He was a largely self-taught naturalist and paid for his expeditions by selling the specimens he had collected, first in the Amazon with Henry Walter Bates and then on his own in the Malay Archipelago. One of the inspirations for his expeditions had been reading Darwin's account of the *Beagle* voyage and the two had met briefly before he had set off for the Far East. They had written to each other on a number of occasions, their correspondence mostly involving Darwin requesting information from Wallace in the same way he requested it from numerous other contacts around the world. Wallace was aware of Darwin's field of interest, which was why he sent him the essay, together with a request that, if Darwin considered the essay to be of any merit, he should forward it to Charles Lyell.

It appears now to have been an extraordinary coincidence that Wallace chose to send the essay to Darwin but, at the time, evolution was a controversial subject which was not accepted by the majority of mainstream academics. Wallace had exchanged a few letters with Darwin touching on the subject and knew that Darwin was sympathetic to the idea. He also knew that Darwin was one of the most respected naturalists in Britain and was socially very well connected, so, if Darwin approved of the essay, it would find a much wider acceptance than if Wallace had sent it directly to a scientific society with the hope of its being published in a journal. As far as Wallace was concerned, he was sending his essay to one of the few men in Britain who could appreciate it and advance his cause. It was very much the case of the younger man seeking approval from somebody he admired. Little did he know the traumatic effect it would have on Darwin.

The content of the essay was a bombshell for Darwin but, ever the honourable Victorian gentleman, he did as Wallace requested and sent the essay on to Charles Lyell, who had previously advised Darwin to publish his work on the subject before somebody else beat him to it. He included a letter of his own with the essay, expressing his anguish to Lyell:

> Your words have come true with a vengeance that I should be forestalled. You said this when I explained to you here very briefly my views on "Natural Selection" depending on the struggle for existence. I never saw a more striking coincidence. If Wallace had my M.S. sketch written out in 1842 he could not have made a better abstract. Even his terms now stand as Heads to my Chapters.[2]

A little later in the letter he goes on to say 'So all my

originality, whatever it may amount to, will be smashed'. This was what he was most worried about. In scientific circles, publishing first on a subject could not only lead to great esteem for the author, but was considered to give them an intellectual priority over the area. Darwin was in danger of losing this priority over the work to which he had devoted his life.

The date on the letter was 18 June and Darwin told Lyell he had received the essay from Wallace that morning. There has been some suggestion that he actually got the letter from Wallace up to a month earlier, a theory based on the timetables of ships travelling between Singapore and England, and that he used the time he had before writing to Lyell to incorporate some of Wallace's ideas into his own work. Wallace appears to have posted his letter some time during the middle of March, which means it is possible for it to have reached England by May. However, Darwin's private writings and actions immediately prior to his contacting Lyell on 18 June show no sign that he was aware of Wallace's theory, so there is no actual evidence to back up the claim that he kept Wallace's essay for a number of weeks before taking any action himself.

Initially, Darwin proposed to give up any claim he had on the idea of evolution but, in an exchange of letters between himself, Lyell and Joseph Hooker, the director of Kew Gardens, he was persuaded otherwise. With the health of his children fully occupying his mind, Darwin left his two friends to decide what to do. They arrived at a solution which they thought would be fair to both Darwin and Wallace. They proposed that a joint paper should be put together, comprising of Wallace's essay and a number of manuscripts by Darwin, including an essay on natural selection he had written in 1844 and a letter outlining his ideas which he had written the previous year to Asa Gray, the American botanist. Both Lyell and Hooker were well connected at the Linnean Society, an institution that they thought was more open to new ideas than some of the more austere scientific societies of the day, and this enabled them to introduce the joint paper onto the agenda of the last meeting of the society before the summer recess.

Darwin had some misgivings, wondering if he was behaving honourably towards Wallace, who had entrusted him with the essay in the first place, but he went along with the plan. The meeting took place on 1 July, a few days after the funeral of Darwin's youngest son, so Darwin was in no state to attend himself. The secretary of the society read the paper out to the assembled members and, considering the momentous content, it appears to have prompted a muted response. Very little discussion occurred, perhaps because neither Darwin or Wallace was there in person to answer any questions or perhaps because the inclusion of the paper meant that the meeting had been extended and there was little time for the usual discussion between the members.

So, after twenty years, Darwin's great idea was finally in the public domain. One of the reasons he had delayed for so long was the controversial nature of the subject matter and now it had been greeted with hardly a murmur of excitement or debate. In fact, when Thomas Bell, the president of the Linnaean Society, made a speech summing up the activity of the society that year, he said:

> The year which has passed has not, indeed, been marked by any of those striking discoveries which at once revolutionize, so to speak, the department of science on which they bear.[3]

In the aftermath of the meeting, Darwin worried about Wallace's reaction, hoping he would not think he had been railroaded into accepting a situation that was beyond his control. Darwin wrote to him in the Far East to explain the course of events, including in his letter a note from Hooker along the same lines, and then had to wait for what must have seemed an eternity for a reply. If Wallace was at all unhappy at finding himself the co-discoverer of evolution rather than the sole originator of the idea, he did not mention anything to that effect in his reply to Darwin or in any of his subsequent writing on the subject. In fact, he thanked Darwin for his efforts and appeared to be happy to have had his name associated with such a famous naturalist.

With the meeting out of the way, Darwin and his whole family left Downe[4] for the Isle of Wight, both to escape the scarlet fever epidemic, which would claim the lives of six more young children in the immediate area that summer, and to allow Henrietta to recover from her own illness. Darwin needed to recuperate himself after the strain of the previous month but, as soon as he arrived on the island, he began writing an abstract of his 'big book'. Hooker had persuaded him that he needed to publish a fuller account of his theory as soon as possible, rather than continue working on what he hoped would be a comprehensive treatment of the subject. He had already been doing this for a number of years and it could take several more years to complete.

Darwin began to work in earnest on his abstract, which grew steadily in size, although he remained reluctant to think of it becoming anything more substantial. Over the next nine months it continued to expand, eventually becoming *The Origin of Species*. Darwin would never actually go on to write his 'big book' on natural selection, although much of the material that did not make it into the *Origin* would be used in his later writings, particularly *The Variation of Animal and Plants Under Domestication* and *The Descent of Man*.

Towards the end of summer of 1858, when the threat from scarlet fever had reduced, the Darwin family returned to Down House and Darwin adopted a strict regime of continuous writing in his study. A combination of factors came together to concentrate his mind on this writing, chief among these the shock he had received with the arrival of Wallace's letter. His theory of natural selection was also now in the public domain, so he needed to make sure of his priority, as Hooker had told him. In addition, he could no longer use the excuse of holding back because of the controversy it would cause. Perhaps the sickness in the family, and the fact that his oldest sister Marianne had also died during the summer at the age of sixty, had made him more aware than ever of his own mortality. It must have become apparent to him that, if he was ever going to get on with writing up his theory, the moment to do so had come.

The writing continued into the following year, with Darwin experiencing periods of elation and depression, as many writers do, depending on how well or badly it was going. In February, the various symptoms of his illness – particularly chronic fatigue – returned and he left Down House for Moor Park, a hydrotherapy centre in Surrey run by Dr Edward Lane. Lane was experiencing his own trials at the time over an accusation of sexual impropriety with a female patient which had ended up in court. Darwin had visited Moor Park on a number of occasions and found the treatment, which, in essence, involved taking long baths, to be one of the few things to ease his symptoms and he was prepared to stick with Dr Lane despite the ongoing sex scandal. Before attending this establishment in Surrey Darwin had been a regular visitor to the spa town of Malvern in Worcestershire for treatment but, after his daughter Annie had died there in 1851, he could no longer face the prospect.

On 12 February, while he was still away from his family at Moor Park, Darwin turned fifty. He doesn't appear to have marked the occasion in any particular way, writing on the day to his cousin and close friend William Darwin Fox, who had been at Cambridge with him, without mentioning its significance, instead describing his illness and saying he thought it had been brought on by his heavy work load. The following day he wrote to his oldest son William, then in his first year at Cambridge and living in the same rooms his father had once had. Again Darwin did not mention his birthday, but he did tell his son about the games of billiards he had been playing. He obviously enjoyed the game, although he claimed to be be a poor player and, on his return home, he ordered an expensive table for the house.

Darwin got back to work again after the trip to Moor Park, keeping up the same relentless pace, although he now took regular breaks to play billiards since he found it relaxing and it took his mind off his work. By May the manuscript was finished. It extended to more than 500 pages and 150,000 words but, even so, Darwin was still thinking of it in as an 'abstract'. However, it was certainly far too long to be published in a journal, as had been the original intention. It had now extended to a sufficient length to be considered a good-sized book in its own right. Lyell suggested to Darwin that John Murray, who published Lyell's own work as well as an edition of Darwin's *Journal of Researches* and who already had a prestigious list of scientific titles, would be a good choice as a publisher. Murray was very well established and well known in the book trade, his father having published the works of Lord Byron, amongst many others. When Darwin and Lyell approached him, he agreed in principle to publishing Darwin's work before he had read it.

The only significant change Murray wanted Darwin to make was to the title. Darwin had suggested *An Abstract of an Essay on the Origin of Species and Varieties Through Natural Selection*, which Murray thought was too long and involved and would put off potential readers. He was, in the end, a commercial publisher who wanted to sell as many copies as possible, so the two men settled on the shorter version of *On the Origin of Species by Means of Natural Selection*, although the title page in the first edition would include the alternative title, *The Preservation of Favoured Races in the Struggle for Life*. On the spine of the book, when it subsequently appeared in a handsome but quite plain green cloth, it simply said *Origin of Species*, with the author's name given immediately below it as Darwin. Such a simple design for the book was no accident, but was done intentionally to make a statement. This was a straightforward scientific volume that was not in any need of any elaborate decoration and, unlike an earlier and anonymously published book on the controversial subject of evolution, this book had the name of one of the most highly respected naturalists of the day on the title page. Everything about the book was serious and scholarly, demonstrating to the outside world that the content of the book could speak for itself.

Darwin, with the help and encouragement of his wife, spent the summer preoccupied with what he found to be the onerous task of going through the proofs. Although John Murray was too polite to mention it, he must have been driven to distraction by the continuing flow of revisions and corrections coming from Down House, each of which would have added to the costs of publication. By 1 October, Darwin had finally finished, much to the relief of all concerned, and the following day he set out for the hydrotherapy centre at Ilkley in the Yorkshire Moors. His health had certainly deteriorated during this intense period of work and worry but, knowing full well the likely reception the book would receive from critics when it came out, he may also have gone to Yorkshire to put more distance between himself and London than he would have done if he had returned to Moor Park in Surrey.

The book was officially published on 24 November 1859 and Darwin, still in Yorkshire taking the water cure, continued his habit of being absent whenever anything important was about to happen in his life. John Murray had decided to print 1,250 copies of the first edition, a relatively modest number which was over-subscribed on the publishing day, allowing him to tell Darwin it had sold out. This was a common low-risk strategy for a publisher to adopt at the time. Rather than print a large number of copies of a book which was, to some extent, an unknown quantity, and risk ending up with lots of unsold copies, many publishers opted to print a small number at first and, if they sold well, to quickly print more for a second edition. This is one of the reasons why a first edition is so expensive today. As well as being an historically important book, a first edition of the *Origin* is quite rare, fetching in the region of £80,000 to £100,000 for a fine copy.[5]

And there it was. After all those years of uncertainty and dithering, after all of Darwin's problems with ill health and the worry about the controversy his theory would cause, the book was finally out. His great idea, that had been on his mind for twenty years and so far had only found expression in the pages of a learned journal, where it had provoked a very modest response, was now out in the open. Anybody with fourteen shillings to spend on a copy, or who wanted to borrow it from one of the burgeoning number of libraries of the time, could read for themselves about natural selection. As he returned to Down House towards the end of November, all Darwin could do was wait for the reaction, which he surely must have known would come.

On the *Origin*

The shelf life of a science book can often be quite short, with new developments in the field moving the subject on and thus making old theories redundant. If an old science book retains any interest at all, it is usually only from a historical perspective. Even those memorable in the history of science can quickly fade into obscurity. *The Origin of Species* is one of the few exceptions, remaining relevant even 150 years after it was first published and retaining its place as the foundation text of modern evolutionary biology. Any serious student of the subject should at the very least be aware of its contents, even if they have not read the book in its entirety. This is not to say that Darwin was right about everything he included in the book. Some of his ideas have been superseded or developed to the point where they are now hardly recognisable, while, on a few occasions, Darwin was just plain wrong. Nonetheless no current work on evolution can take place without reference being made to him in some form.

The *Origin* is also not a typical science book in that it does not contain any tables or statistics and there is only one diagram, illustrating what Darwin describes as 'the

Tree of Life'. Darwin's writing style obviously belongs to the Victorian period and it can occasionally be a little long-winded, but it is fundamentally engaging. Rather than simply containing dry lists of facts, as so many science books do, it is often personal and anecdotal. What the *Origin* does is to present an argument, putting the case for natural selection, and then goes on to examine and explain the reasons why this argument is valid.

Later in life Darwin would say he could only read light romantic novels, preferably ones with a pretty heroine and a happy ending, but he was also well acquainted with what we might now think of as the classic authors of the period, such as Charles Dickens and George Eliot, as well as with those of earlier centuries such as Shakespeare and Milton[6]. While his own work was hardly in the same league as these authors, he was still an accomplished writer. Since his return from the *Beagle* voyage, one of his primary activities had been writing and, by the time he had begun work on the *Origin*, he had twenty years of experience behind him.

Darwin's writing style may make the book reasonably easy to read, certainly when compared to most science writing, but it is hardly the main reason to read the book. It is the content that has given it an enduring quality but Darwin expresses what can be difficult concepts in a fluent and straightforward manner. In the opening sentence of the final chapter, which is a summary and conclusion to the book, Darwin describes it as being 'one long argument'. The argument he is referring to is, in essence, a defence of his theory of descent with variation

– what we now call evolution. Although he would include it more frequently in later editions, Darwin used the word 'evolution' very sparingly in the first edition of the book, most probably because, at the time, the word had become associated with radical, even revolutionary, ideas. Darwin thought of himself as a scientist first and, while he cannot possibly have been unaware of the controversial nature of what he was proposing, he was doing so firmly from within the scientific establishment.[7]

The main theme of the argument was that variation within members of the same species actually occurred and the process by which they varied was natural selection. Darwin gave a succinct explanation of what he meant by natural selection in the introduction:

> As many more individuals of each species are born than can possibly survive; and as, consequently, there is a frequently occurring struggle for existence, it follows that any being, if it vary however slightly in any manner profitable to itself, under the complex and sometimes varying conditions of life, will have a better chance of surviving, and thus be naturally selected.

Here, in a single, if rather long, sentence, is the core of the matter. Not only does variation occur, Darwin is saying, but this is the process by which it occurs. Those individuals of a species of an animal or plant best adapted to their environment are those which will be the most successful and will have more offspring than less well adapted individuals. Over time, successive generations of these successful individuals will come to dominate the population of the species and, if that population is isolated from other populations of the same species, it will begin to diverge from them and may eventually form an entirely new species.

On the face of it, this a a relatively simple idea, but the implications, and what can be inferred from the implications, are enormous. For example, if new species can arise through a natural process rather than being the creation of God, then the logical extension of this idea is that there is no reason to suppose that a Creator exists at all. Darwin did not take his argument down this road, but it is a simple enough path for anybody reading the *Origin* to follow. This is, of course, why the book was so controversial.

The occurrence of evolution by natural selection is now accepted by the overwhelming majority of scientists, although the details are still rigorously debated, but the opinion of the scientific establishment during Darwin's day was very different. Science was dominated by the views of the Anglican Church, particularly at Oxford and Cambridge, which held to the position of the immutability of species, the idea that each species was created by God and did not vary over time, with each successive generation being much the same as the previous one. A challenge to this orthodoxy was considered to be not only a challenge to the scientific establishment, but also to the order of society.

So, right from the introduction of the book, Darwin was making his position very clear. Of course, having set out his argument in the introduction, it would not be unreasonable to wonder what Darwin's next 500 pages were about. What he does is carefully to assemble the argument in favour of natural selection, backing it up with numerous examples, and then critically examine it, explaining the problems and anticipating any criticism before it has arisen.

The main body of the text begins with a discussion of the breeding of a number of varieties of domestic animals and plants, including pigeons, which Darwin bred himself at Down House. This may appear to be a strange place to begin a book on the natural world, but Darwin had a particular purpose in mind. Rather than beginning with examples of variation that very few people in Victorian England would have been familiar with, such as the now famous Galapagos finches, he chose to illustrate his argument with animals and plants known to everyone: dogs, cows, sheep and, in particular, pigeons. In a long section he describes how all the various domestic varieties of pigeon – the carriers, tumbles, fantails and the rest – are all descended from the wild rock pigeon. An experienced breeder could select for various traits, producing a new variety in only a few generations which, Darwin says, somebody unfamiliar with pigeons in general would take for an entirely different species from the variety from which it was bred.

Having introduced the subject of variation between individuals of the same domestic species in a form which is hard to argue against, Darwin moves on to variation in nature, providing plenty of examples of the variety within species across their geographical ranges. He then describes what he calls 'the struggle for existence' – how most individuals of a species produce a great deal more offspring than can possibly survive, which then compete with one another for the available resources – and says that he has adapted this idea from the work of the political economist Thomas Malthus (1766–1834) on human population growth.

Natural selection, Darwin explains, is an extension of Malthusian ideas of population growth as applied to the natural world. He used the term 'natural selection' to contrast it to domestic selection, although he would later acknowledge that the terminology could lead to some confusion. The use of the word 'selection' implies some outside agency is doing the selecting, as plant and animal breeders do in domestic selection. However, what Darwin is actually saying is that the variation caused by natural selection is an entirely natural process, not governed or directed by anything other than the prevailing environmental conditions and competition for resources.

The next step in the long argument was to put the struggle for existence together with variation over generations. The result of this, Darwin says, is natural selection, a process he describes as 'the preservation of favourable variations and the rejection of injurious variations'. He goes on:

> It may be said that natural selection is daily and hourly scrutinising, throughout the world, every variation, even the slightest; rejecting that which is bad, preserving and adding up all that is good; silently and insensibly working, whenever

> and wherever opportunity offers, at the improvement of each organic being in relation to its organic and inorganic conditions of life. We see nothing of these slow changes in progress, until the hand of time has marked the long lapse of ages, and then so imperfect is our view into long past geological ages, that we only see that the forms of life are now different from what they formerly were.

Much of the rest of the book is given over to providing illustrations of natural selection, explaining particular points in more detail and describing several problems with the theory in anticipation of likely criticisms. Many of the examples he uses come from oceanic islands, particularly those with which Darwin was personally familiar, having visited them on board the *Beagle*. He shows how species of animals and plants which are unique to an island are, in fact, related to and descended from other species on the continental mainlands.

One point Darwin is particularly keen to make is that evolution is a gradual process and he builds on the geological work of Charles Lyell, which demonstrated that the Earth, far from being a few thousand years old, as had been proposed by some Biblical scholars, who arrived at a figure by calculating how long it would have taken for the events described in the Old Testament to have occurred,[8] is actually many millions of years old. Darwin describes natural selection as being an ongoing but very slow process, requiring very long periods of time to produce the great variety of life on Earth. This length of time, as Lyell showed, is accounted for in the geological record.

We now know, of course, that the planet is much older than the hundreds of millions of years envisaged by Lyell and Darwin. The Earth is actually something like 4.5 billion years old and the first life forms evolved about 3.5 billion years ago.

Darwin draws all these disparate strands together in the final chapter and the book ends with what would become one of its best known passages, the opening sentence of which shows that Darwin was well acquainted with Shakespeare in its echoes of a scene from *A Midsummer Night's Dream*[9]:

> It is interesting to contemplate an entangled bank, clothed with many plants of many kinds, with birds singing on the bushes, with various insects flitting about, and with worms crawling through the damp earth, and to reflect that these elaborately constructed forms, so different from each other, and dependent on each other in so complex a manner, have all been produced by laws acting around us.

The book concludes with one of the most often quoted lines in science:

> There is a grandeur in this view of life, with its several powers, having originally breathed into a few forms or into one; and that, whilst this planet has gone cycling on according to the fixed law of gravity, from so simple a beginning endless forms most beautiful and most wonderful have been, and are being, evolved.

The eloquence of Darwin's writing about the beauty of

the world has rarely been matched. In describing the planet as 'cycling', he is alluding to Copernicus and he then makes a more direct reference to Newton and the law of gravity, thereby placing natural selection firmly within the scientific tradition of explaining how the world around us works by empirical means.

In the 150 years since the publication of the *Origin* many of the problems with evolution which Darwin himself identified have been solved, including the question of how complex organs such as the eye have evolved. (Even so, this example is one still raised by opponents of evolution to justify their position.) The development of the science of genetics, beginning in the early twentieth century and based on the rediscovery of the work of Gregor Mendel, has unravelled the mechanism by which inheritance occurs and has also shown how variation can be generated and passed on through the transmission of genes. The lack of this knowledge of the mechanism of inheritance was a major stumbling block for Darwin. It is somewhat ironic that Mendel was conducting his experiments on breeding peas at the same time as Darwin was writing the *Origin*, and yet the two fields of knowledge would not become integrated for another eighty years in what is known as the 'modern synthesis'.

The Controversy Begins

The publication of *The Origin of Species* was not only a landmark occasion because of the revolutionary affect it would have on the natural sciences. It became the focal point of a national debate, which would go much further than the contents of the book itself. It was perhaps the first time such a debate on a large scale had occurred in Britain. Over the previous hundred years, the industrial revolution had led to fundamental changes in British society and the rapidly expanding population was increasingly keen to question the established order.

By the 1850s Britain had emerged from a period of political and social unrest which had begun in the early 1830s. Mass movements formed by the middle and lower classes had demanded change, bringing the country to the brink of revolution on a number of occasions. Revolutionary change had only been avoided by concessions to their demands in the shape of political and social reforms and because a booming economy had led to an increasingly affluent society. The main beneficiaries of this wealth creation were the burgeoning middle classes, with money generated through commerce, trade and manufacturing and through a rapidly expanding empire. With the Royal Navy ruling the waves and London being the major financial and commercial centre of the world, Britain was, in effect, the only superpower, and no other country was capable of challenging its supremacy.

Darwin's family actually provide a good example of the development of this middle class. One side of the family, the Darwins, came from the professional classes. Both Darwin's father and grandfather were doctors and much of the wealth of the family came from his father's investments in the financial markets. Josiah Wedgwood, Darwin's other grandfather, was a self-made man, amassing a fortune from manufacturing in the pottery business.

The *Origin*, then, emerged into this dynamic and rapidly changing society. Its publication coincided with a boom in the publishing and media industries, along with a rapidly improving postal service and expanding rail network. Access to education was widening and the availability of information throughout society was increasing. This was a result not only of the expanding media industry, with its newspapers and periodicals, but also because of the development of lending libraries and the growing number of public meetings and lectures. Science and technology were an integral part of this burgeoning thirst for knowledge. The Great Exhibition held in London in 1851 as a showcase of industrial technology, for example, attracted six million visitors. By any standard, this is a very large number of people but, when it is remembered that the entire population of the country was around seventeen million at the time, it becomes truly extraordinary.

The audience for serious scientific work was there and the means of distributing new ideas were in place but, in the end, what caused the ensuing controversy was the contents of the book itself. Darwin was well aware of the likely reaction because he knew that the publication of a book on evolution in 1844, while he had been in the process of developing and writing up his own ideas on the subject, had caused a sensation. The book, *Vestiges of the Natural History of Creation*, was published anonymously, although it is now known to have been written by the Scottish journalist and publisher Robert Chambers. It dealt with the origins of life and human evolution, as well as the development of the variety of life, and it was these two areas that provoked the bitterest attacks from the scientific and religious establishment. The nature of the storm of criticism aimed at *Vestiges* was one reason why Darwin not only delayed the publication of his own work but also, despite the title of his book, did not actually deal much with the origin of life when he did finally publish it. He made only the occasional oblique reference to human evolution in the *Origin*, although he would go on to deal with the subject fully in *The Descent of Man*, which was published in 1871.

If Darwin thought he would lessen the critical response to *Origin* by omitting the most controversial aspects of evolution, he was wrong. The subject was already a part of public awareness and, despite Darwin's attempts to lessen the controversial nature of the material, connecting his theory with previous work on evolution was an obvious step to take. *Vestiges* had essentially been speculative in nature, which allowed it to be easily dismissed because of lack of evidence. Darwin himself had done so. The *Origin* not only presented an impressive body of evidence but, in natural selection, a mechanism by which evolution occurred. The other major difference between the two books was that the author of the *Origin* was in no sense anonymous. Darwin's name was on the title page of the book and, as he was one of the best known and respected naturalists in Britain at the time, his words could not be so easily dismissed. Opponents of evolution could not simply ignore the book; they had to refute its central argument.

The response was quick in coming, and both positive and negative reviews began to appear in newspapers and periodicals in a matter of a few days after publication. Letters began to appear in *The Times*, then very much the mouthpiece of the British establishment. One of the letters was by Robert FitzRoy, Darwin's captain on the *Beagle*. Although written under a pseudonym, the identity of the author was immediately apparent to Darwin, who was highly dismissive of the views of his former captain in a letter to Charles Lyell.

One of the surprising aspects of the criticisms initially levelled at Darwin, at least from a modern perspective, is that it came largely from what we would call a scientific standpoint rather than a religious one – although, at that time, it was not always easy to distinguish between the two. A literal interpretation of the Bible, including the account of the Creation given in Genesis, was not as prevalent in nineteenth century Britain as it would become in twentieth and twenty-first century America. The debate about the creation of life was carried out between scientists, while the religious aspects of the argument were part of a wider debate about the role of religion in society in general. Those who argued in favour of Darwin's position also tended to argue in favour of a more secular society, where religion played a reduced role in all sectors of society as well as in science.

Looked at in the simplest way, the two sides opposing one another in the debate can be characterised as, on the one side, established scientific figures who had built their careers on the orthodox view of creation and, on the other, a younger generation who were more open to new ideas and to change, although the reality was rather more complex, since such established figures as Charles Lyell were in favour of evolution.

One opponent of evolution who was a pillar of the establishment was Adam Sedgwick, who was one of Darwin's mentors from his Cambridge days and who had played an important part in introducing him to geology. He combined his role as a Cambridge professor of geology with that of a clergyman and was firmly on the creationist side of the argument. After reading the *Origin* shortly after it was published, he wrote to Darwin to express his dismay at the direction Darwin's thinking had taken, saying:

> I have read your book with more pain than pleasure. Parts of it I admired greatly; parts I laughed at till my sides were almost sore; other parts I read with absolute sorrow; because I think them utterly false and grievously mischievous.

At the end of the letter Sedgwick told Darwin that, despite their 'disagreement in some points of the deepest moral interest', he remained Darwin's 'true-hearted old friend'. Darwin replied a few days later in the same spirit, maintaining his views but, at the same time, showing an unwillingness to fall out with an old friend:

> You could not possibly have paid me a more honourable compliment than in expressing freely your strong disapprobation of my book – I fully expected it. I can only say that I have worked like a slave on the subject for above 20 years

> and am not concious that bad motives have influenced the conclusions at which I have arrived. I grieve to have shocked a man I sincerely honour. But I do not think you would wish anyone to conceal the results at which he has arrived after he has worked according to the best ability which may be in him.

Needless to say, not all of the debate was conducted with such a regard for politeness. One of the more vociferous of Darwin's critics was Richard Owen, the superintendent of the natural history collections of the British Museum, who would later be instrumental in founding the Natural History Museum in London. By all accounts Owen was a difficult man to like. He was not above advancing himself at someone else's expense and had been accused of claiming the credit for work done by other people. At the same time, he was one of the foremost comparative anatomists in Britain, particularly in the study of fossils, and he is credited with inventing the word 'dinosaur'. Darwin knew Owen well, having been introduced to him by Lyell shortly after returning to Britain from the *Beagle* voyage. Owen had worked on the anatomical description of some of the fossils Darwin had brought back from South America and they had kept in touch in the years since.

After reading the *Origin,* Owen wrote an anonymous review for the *Edinburgh Review* which personally attacked both Darwin and his supporters and described Darwin's work as an 'abuse of science'. The identity of the author was immediately apparent to Darwin and his supporters and a long running feud broke out between them, one mostly carried on by the supporters rather than by Darwin, since he himself was intent on maintaining his usual low profile. Some years later it emerged that Owen had secretly been campaigning to block Government support of the Botanic Gardens at Kew, which was run by Joseph Hooker, one of Darwin's oldest friends and staunchest supporters. It was the last straw as far as Darwin was concerned. In the autobiographical sketches he wrote towards the end of his life he was uncharacteristically scornful of Owen, writing:

> I often saw Owen, whilst living in London, and admired him greatly, but was never able to understand his character and never became intimate with him. After the publication of the *Origin of Species* he became my bitter enemy, not owing to any quarrel between us, but as far as I could judge out of jealousy at its success.

With eminent men like Owen ranged against him, it is just as well for the retiring Darwin that he had such an able group of supporters. As well as Hooker, Charles Lyell came out in support of Darwin, although he could not bring himself to agree with all of Darwin's conclusions, remarking on one occasion that he 'could not go the whole orang'. In America, Asa Gray vigorously defended evolution against his fellow Harvard professor Louis Agassiz, the foremost American naturalist of his day and a firm believer in the divine nature of creation, even though Gray, a devoutly religious man himself, also could not accept all of the implications of evolution.

The one man to promote the cause of evolution and to defend Darwin publicly at every possible opportunity – perhaps more than any other – was Thomas Henry Huxley. In some respects Huxley can be compared to Richard Dawkins, a modern day public defender of Darwinian evolution. He was tenacious and unyielding in debates, both in print and at public meetings, and was often deliberately provocative in attempts to goad his opponents. In this respect he was very different from Darwin, but there were also some striking similarities. He was from a very different and far less wealthy background than Darwin, so he had always had to earn a living for himself and his family but, as a young man in his twenties, after some years training in medicine, he had embarked on board HMS *Rattlesnake* on a surveying expedition which would last four years, much as Darwin had set off on the *Beagle*. But, whereas Darwin had travelled on the *Beagle* as a companion to the captain and had paid his own way – or, at least, his father had paid for him – Huxley had gone as the surgeon's assistant. Part of his duties included helping with scientific experiments and collecting natural history specimens and, again like Darwin fifteen years earlier, he had gained a reputation as a naturalist by the time he returned to Britain.

Darwin and Huxley knew each other well. They were part of the same circle of naturalists and scientists who shared similar interests, which also included Hooker and Lyell, and they corresponded regularly over a period of years as well as dining together during Darwin's trips into London or when Huxley visited him at Down House.

After reading the *Origin* within a few days of its publication, Huxley told Darwin in a letter, 'no work of Natural History Science I have met with has made so great an impression on me' and went on to say that he felt 'endowed with an amount of combativeness' for the fight he knew was coming.

Even before the publication of the *Origin*, Huxley had locked horns with Owen. In the previous year he had shown that Owen's assertion that certain anatomical differences between the brains of human beings and those of gorillas indicated that there was not a close relationship between them was entirely wrong[10]. There appears to have been a great deal of personal animosity between the two men and Huxley relished the opportunity afforded to him by the publication of the *Origin* to renew the battle with Owen. It also allowed him to continue a wider argument he was also already engaged in over what he perceived as the Anglican Church's stranglehold over science which, he considered, had been holding back his career. He wrote a number of reviews of the book, making crystal clear in all of them his support for Darwin and his disdain for those who opposed him. In the review published in the *Westminster Review* he wrote:

> Extinguished theologians lie about the cradle of every science as the strangled snakes beside that of Hercules, and history records that whenever science and dogmatism have been fairly opposed, the latter has been forced to retire from the lists, bleeding and crushed, if not annihilated; scotched if not slain.

Huxley could be as forthright in public debate as he was in print. At a meeting of the British Association for the Advancement of Science held in Oxford during June 1860 Huxley took part in a debate which would become one of the most famous occasions in the history of science. Darwin had said he would attend the meeting himself but, a few days before it took place, he suffered yet another recurrence of his illness, preventing him from going and leaving the floor to Huxley.

Accounts of the debate itself differ widely and, as no record was kept of what was actually said, it is difficult to say now what really happened in its entirety. However, it seems clear that Samuel Wilberforce, the Bishop of Oxford and son of William Wilberforce, the campaigner for the abolition of slavery, spoke against evolution and asked Huxley at the end of his speech if he was related to an ape on his grandfather's or his grandmother's side. Recalling the events much later, Huxley wrote that, as soon as the Bishop had made this throwaway quip, he whispered to the man sitting next to him, 'The Lord hath delivered him into mine hands', before getting to his feet and saying words to this effect:

> If the question is put to me would I rather have a miserable ape for a grandfather or a man highly endowed by nature and possessed of great means of influence and yet who employs these faculties for the mere purpose of introducing ridicule into a graven scientific discussion, I unhesitatingly affirm my preference for the ape.

The audience apparently cheered and jeered in equal

measure, some people applauding Huxley and others expressing shock that he could address a Bishop in such a manner. A number of other people spoke after Huxley, including Hooker, and Robert FitzRoy, who was in the audience, is said to have jumped to his feet, waved a Bible above his head and denounced Darwin, regretting he had given him the opportunity to collect the facts he would go on to use in such a disgraceful fashion, before he was shouted down by other members of the audience. The debate broke up with both sides claiming they had won the argument, but the legacy of the meeting was to show that it was no longer necessary to defer to the authority of senior members of the Anglican Church in matters of science and, in this respect at least, Huxley had come out very much on top.

In succeeding years the acrimonious tone of the debate died down and the idea of evolution became less controversial. The scientific and religious establishment began a process of reaching an accommodation with the ideas of evolution. Herbert Spencer's term 'the survival of the fittest' could easily be adapted to fit with the prevailing ethos of free market economics and *laissez-faire* capitalism, allowing the upper echelons of society to argue that they had risen to their positions because they were the 'fittest' people in society and so had achieved their privileges through a natural process. In this way, evolution could be co-opted to maintain the established order of society rather than to threaten it.

Darwin expressed a certain amount of sympathy with these ideas – he was, after all, a Victorian gentleman himself – but, in his subsequent writings, he chose to concentrate on the areas he knew best, natural history and what would come to be called evolutionary biology. His cousin, Francis Galton, was not so reticent, developing a theory of social evolution which he called eugenics. At its heart, this involved the idea that human society could be improved by what amounted to selective breeding and it would, in the twentieth century, be used as a justification for any number of racist ideologies and discriminations against minorities.

Since Darwin's day evolution has become a central tenet of the biological sciences, accepted by almost everyone in the scientific community. This is not to say that there are no longer any arguments between scientists, because there certainly are, but these arguments concern different interpretations of how evolution occurs, not if it occurs at all. The controversy, however, has not gone away. Today it is being continued by a different group of people. The emergence of Christian fundamentalism, particularly in America, has seen the argument against evolution taken up by religious groups who believe in a literal interpretation of the Bible. As there is no way of reconciling evolution with the Biblical creation in Genesis if the latter is taken as a literal account, there does not appear to be any way of resolving this dispute.

In an attempt to take the argument back into the realms of science, a version of creationist thought called intelligent design has been developed. This is essentially an updated version of the teleological argument, which suggests that the complexity of life on earth cannot possibly be the result of a chance natural process, no matter how long that process has been going on, so it must be evidence of concious design at work in nature and, by extension, of the existence of God. This has been presented as a scientific counter-argument to evolution, even though it relies on faith rather than rationality. Thomas Huxley, if he were alive today, would be having a field day combating an idea that can be refuted so easily.

The Philosophical Naturalist

As the initial fury of the religious debate began to die down, a number of other criticisms of Darwin began to find expression, including some of a more personal nature than those which simply opposed the ideas he set out in *The Origin of Species*. One of these, a criticism that has resurfaced again and again over the years, accused Darwin of simply being a wealthy amateur naturalist who either adopted other people's evolutionary ideas, such as those of Lamarck, for his own purposes, or accidentally stumbled across natural selection without really knowing what he was doing.

The first of these accusations arises from a misunderstanding of what Darwin was trying to achieve. He did not invent evolution, which was a current if not widely accepted idea before Darwin began working on his theories of species change. The question to which Darwin was trying to find an answer was not whether or not species could change, but how they changed. Lamarck had developed the most coherent theory to explain this before

Darwin, with his hypothesis of the inheritance of acquired characteristics, which suggested that changes in the physiology of an organism occurring during its lifetime could be passed on to future generations. Darwin was certainly aware of these evolutionary ideas, but dismissed them because they did not explain how these changes could happen. What Darwin succeeded in doing was finding the mechanism of change – natural selection – so that, rather than simply speculate that change *could* happen, he was able to show *how* it could happen.

The second charge – that Darwin simply stumbled across natural selection – is curiously similar to the idea that he was a genius who plucked the fully formed theory out of thin air in a moment of inspiration. In both scenarios, one of which casts him as stupid and the other which has him possessing some sort of otherworldly intelligence, he is depicted as the conduit of an idea without really being aware of what he was doing. Both accusations fail to recognise the years of training as a geologist and naturalist he went through on board the *Beagle* and after he had returned to England, and also the extraordinary amount of sheer hard work he put in before reaching his conclusions.

Throughout his life, Darwin kept extensive records of his work, much of which survives and has been published in book form or online. From these records it is possible to trace his development as a practical and theoretical evolutionary biologist, as he would now be known, giving the lie to the idea that he developed his theories by accident. Parts of these records include lists of the books he read and the notes he kept on those he thought would be important or useful for his work. The breadth of his reading revealed in these lists is astonishing, ranging from the expected works in the fields of natural history and geology to a wide spectrum of books in other disciplines, particularly philosophy and political thought.

On a number of occasions Darwin described himself as a philosophical naturalist and he probably meant that he did not just observe the natural world and blindly catalogue it, but thought about what he saw. But he could equally well have been referring to the fact that his writing can be seen as fitting into the Western philosophical tradition of examining the nature of existence through the use of critical reasoning. This is not to say that Darwin was engaged in writing works of philosophy, rather that he was aware of the intellectual progression in the field. At different times in his writing he referred to some of the great names of European philosophy, including Descartes, Leibniz and Kant, as well as Bacon, Hobbes and Locke from the British tradition.

Although he did not make specific reference to them – one of Darwin's habits was that he often did not attribute his sources properly – it can be inferred that he was particularly influenced by two philosophers from what is sometimes called the Scottish Enlightenment, an eighteenth-century intellectual flowering centred on Edinburgh. The first of these was David Hume, who promoted the philosophy of empiricism, which asserts that knowledge can only arise from experience and observation rather than from any supernatural or divine source.

He expressed his views as a form of extreme scepticism, showing that explanations of the natural world did not depend on religious foundations but on observable laws. The rigour with which Hume expressed these views meant that, at least for anybody who chose not to ignore them, it was no longer possible to accept religious explanations for the natural world without questioning them. Darwin did precisely this after he returned from the *Beagle* voyage.

The other important figure from the Scottish Enlightenment was Hume's contemporary and close friend Adam Smith, now mostly remembered as the father of modern free market economics, although his writing showed he was concerned with many other issues in moral philosophy and political thought. The essence of Smith's work that is applicable to Darwin is the stress he puts on the role of the individual in society. Smith thought that economic order was not achieved by imposing laws on a commercial system, but would arise by allowing the individual to struggle for personal benefit through unrestricted competition. He used the metaphor of an invisible hand guiding the system to explain that somebody competing for his or her own benefit was not doing so to achieve overall economic stability, but that stability would arise as an unintended consequence of competition between all the individuals making up the society. An unrestricted market will, according to Smith, favour the most efficient enterprise, which will become more successful than a wasteful business. In the same way, according to Darwinian natural selection, the individual best adapted to its environment in a population of animals or plants will have the most breeding success.

It is difficult to ascertain the exact extent to which Darwin drew on Hume and Smith and, perhaps, easy in hindsight to find parallels between their work where none existed. It is certainly feasible that Darwin arrived at his conclusions without referring to the work of either man, although he did acknowledge the debt he owed to Thomas Malthus, whose work was highly influenced by Smith. But, whatever the truth of the matter, the commonality of ideas expressed by all three, and in particular by both Smith and Darwin who saw that competition between individuals would unknowingly lead to stability, makes it hard to dismiss Darwin as a lucky amateur. Rather, it shows him to have been a rigorous thinker in his own right, who worked within the intellectual climate of his age in an attempt to uncover universal truths about humanity and the world within which we live. The great success he had in this field proves him to have been one of the great thinkers of his age and one that has to be taken into account by modern philosophers working in the fields of the philosophy of the mind or the philosophy of science.

The Making of a Naturalist

Early Years

When considering the life of any well known person, it is always tempting to look for the signs of what is to come in his or her childhood. This is certainly justified in some cases – the example of Mozart comes easily to mind – but, in truth, Charles Darwin's childhood was not particularly remarkable. In a short piece of autobiography written in 1876, he recalled:

> When I left school I was for my age neither high or low in it; and I believe I was considered by my masters and by my father as a very ordinary boy, rather below the common standard in intellect. To my great mortification my father once said to me, "You care for nothing but shooting, dogs, and rat-catching, and you will be a disgrace to yourself and all your family."

There may be a certain amount of Victorian modesty in Darwin's thoughts, and his father's words, although they carried sufficient weight for Darwin to remember them something like fifty years after they were said, are the sort of thing fathers say to their sons when they appear intent on idling their lives away. But they give an indication of a normal and, with a few notable exceptions, mostly happy childhood. Anyone attempting to delve into Darwin's early years in an attempt to uncover the reasons why this particular child would go on to become the greatest naturalist of his day and have his life and work celebrated two hundred years after his birth is likely to come away disappointed.

Charles Robert Darwin was born on 12 February 1809 in Shrewsbury, the county town of Shropshire, and, as a child, was known to his family as Bobby. He was the fifth child of six born to Robert and Susannah Darwin, and had three older sisters, Marianne, Caroline and Susan, who were between six and eleven years older than him, an older brother Erasmus, five years his senior, and a sister, Catherine, who was a year younger. His mother, the daughter of Josiah Wedgwood, suffered from regular bouts of illness when Darwin was very young and died at the age of 52, when Darwin was eight years old. Without doubt it was the most, and quite possibly only, traumatic experience of his childhood. He would later write that he could hardly remember her at all, only the shape of her dresses and going on walks with her, and, while he could remember the funeral of a soldier held in Shrewsbury at about the same time, he had no memory of his mother's funeral at all. From then on his older sisters, particularly Caroline, took over the role of mother in the lives of the younger children, which was a continuation of what they had been doing anyway while Susannah Darwin had been ill. The loss of a mother at such a young age must have had an enormous impact on Charles, as it would on anyone, but it does not appear to have led to any emotional problems either at the time or later in life. Darwin's own frequent bouts of ill health could have had their roots in growing up in a household where illness was an everyday occurrence and became a means of gaining attention but, with no definitive diagnosis of Darwin's illnesses, this can only be a matter of speculation. The overall picture of the family after Susannah's death is of a close and caring one dealing with difficult circumstances as best they could.

Robert Waring Darwin had moved to Shrewsbury in 1786, at the age of twenty, to set up in practice as a doctor, having qualified at the University of Leiden in the Netherlands and taken an MA at Edinburgh University. He was financed in this enterprise by his father Erasmus Darwin (1731–1802), who was also a doctor and was a well-known and well-connected man in Georgian Britain. Erasmus was something of a polymath, being a writer, inventor and natural philosopher, and was also known as a freethinker and nonconformist. One of the many subjects he wrote about was evolution, following on from the thoughts of the pioneering French naturalist, the Comte du Buffon (1707–1788), who would also influence Jean-Baptiste Lamarck. Some of Erasmus Darwin's ideas about evolution were expressed in verse, which perhaps goes some way to explaining why he is less well known in this field than Lamarck is today even though their conclusions were similar. Darwin's unconventional lifestyle, sharing his house with a wife and a mistress, and some of his more radical views, for example his support of the French Revolution, which did not go down well at the beginning of the nineteenth century when Britain was at war with France, have also tended to overshadow his other achievements.

In all probability Robert Darwin shared many of the views held by his father but, with his position in Shropshire society as doctor to the aristocracy, he appears to have kept them very much in the background. The family were Unitarians, attending a chapel in Shrewsbury rather than the Anglican church but, like the Wedgwood family, as their position in society increased, they gradually began to shift away from nonconformity. Darwin was, for example, christened in the Anglican church in Shrewsbury rather than the Unitarian chapel. In the early nineteenth century, when anything radical or unorthodox was viewed with suspicion, it paid for anybody wanting to get on in life to conform.

The connection between the Darwin and Wedgwood families began with Erasmus Darwin and Josiah Wedgwood. Both belonged to the same social and intellectual circles in the English Midlands and met regularly at a number of clubs where they were both members. One of these was the Lunar Society, so called because meetings were held on nights when there was a full moon and members could benefit from the moonlight as they made their ways home after the meetings had finished. Some of the leading intellectual figures in the industrial revolution were members of the club, including Matthew Boulton and James Watt and the natural philosopher and dissenter Joseph Priestley.

At the time the Lunar Society was holding its meetings,

Erasmus Darwin lived in Lichfield, not all that far from the Wedgwood factory at Burslem, one of the six towns of Stoke-on-Trent in Staffordshire. The two families virtually grew up together, and Robert Darwin knew his future wife from a very early age. After they married, Susannah received £25,000 in an inheritance on the death of her father. They invested the money in land and property and built Mount House, a mansion to the north of Shrewsbury, for their expanding family.[11] From then on Robert Darwin would add greatly to the income he gained from his medical practice with a large portfolio of investments and, in a time before loans were available through banks, by lending money to businesses and individuals. By the time of his death in 1848, he had become a very wealthy man, leaving more than £200,000 in his estate, a sum which would today equate to him being a multi-millionaire.

Robert Darwin was a big man, both physically and in personality. He was 6 feet 2 inches tall and, as Darwin would write, stopped weighing himself after he had reached 24 stones. When visiting an unfamiliar house, his son would go on to say, Robert Darwin sent a servant in before him to test that the floorboards were strong enough to hold his weight and he had to have stone steps made to enable him to get into his carriage because he kept breaking the wooden ones. Although he could be over-bearing and expected obedience from his children, Darwin was keen to point out his many good qualities, describing his father as 'a remarkable man'.

The Wedgwood family were frequent visitors to Mount

House and, particularly after Susannah Darwin's death, found the atmosphere stuffy and quite strict. Josiah Wedgwood's son, also called Josiah, took over the family business after the death of his father and bought an estate at Maer, about thirty miles north of Shrewsbury. Darwin was frequently at Maer, where life was more relaxed than in his own home, and it was, no doubt, in the grounds of the estate that he first learned the hunting skills which would attract the disapproval of his father but would later become very useful to him as a naturalist collecting specimens. Uncle Jos, as Charles called Josiah Wedgwood II, would be an important influence on his life and he would ask his uncle (and future father-in-law) to intervene on his behalf when he was attempting to persuade his father about a course of action he wanted to pursue.

Darwin's education began at home, where he was tutored by Caroline. From the age of eight, he attended the day school at his local Unitarian chapel, although he did not excel as a pupil. His interest in natural history, however, was already apparent and he would spend long hours on his own in the countryside, writing in his autobiography:

> By the time I went to this school my taste for natural history, and more especially for collecting, was well developed. I tried to make out the names of plants and collected all sorts of things, shells, seals, franks, coins, and minerals. The passion for collecting, which leads a man to be a systematic naturalist, a virtuoso or a miser, was very strong in me, and was clearly innate, as none of my sisters or brother ever had this taste.

After a year at the day school, Charles became a boarder at the well-respected Shrewsbury School, a public school which was hardly a mile from Mount House. The head-master, Dr Butler, based the education on the classics of Greek and Latin literature, as all public schools did at the time. Charles was interested in natural history and was developing an appreciation for other science subjects, particularly chemistry. When not at school he and his brother Erasmus would conduct their own experiments but such independent activities did not find favour with Dr Butler. Charles did not excel at classical study and, although he appears to have been popular with his fellow students, he did not remember his school days fondly, writing:

> Nothing could have been worse for the development of my mind than Dr. Butler's school, as it was strictly classical, nothing else being taught accept a little ancient geography and history. The school as a means of education to me was simply a blank.

In 1825, when Charles was sixteen, Robert Darwin, aware that his younger son was not thriving at Shrewsbury School, sent him to Edinburgh University, where his older brother had already been for two years, to follow in the family tradition of studying medicine. Although Darwin would not complete the course, it was at Edinburgh that he would start out on the course that would ultimately lead him to becoming a naturalist and writer.

Edinburgh and Cambridge

In the summer of 1825, before going to university, Darwin helped his father in the medical practice, attending patients and writing up notes. Robert Darwin thought that his son's sympathetic and understanding nature would make him a good doctor in his own right and, in October of that year, Darwin travelled up to Edinburgh to begin his medical studies.

In the early nineteenth century, the medical school at Edinburgh had the reputation of being the best in the country. Oxford and Cambridge remained strictly Anglican, not allowing either Catholics or non-conformers to attend at all, but Edinburgh was much more cosmopolitan in outlook, being open to all who could afford to pay. As a result the university was much more receptive to new ideas, including those of a radical nature. The sciences, including chemistry and geology, were very much a part of the curriculum and the classics were not considered essential, at least for medical students. The sixteen-year-old Darwin shared rooms with his older brother who was in the final year of his own medical degree when Charles arrived. But Charles initially did not find the course he had set out on any more congenial than he had previously found Shrewsbury School. The lectures were incredibly dull and the practical aspects of the course revolted him. He attended physiology demonstrations and watched, in the days before any sort of anaesthetic was available, a number of operations. He was deeply distressed by what he saw and knew that a medical career would not be the right one for him.

The social side of university life and time spent both with his brother and with his fellow students was, on the other hand, much more agreeable. Darwin was an easy-going, sociable young man and was intent on enjoying his time at Edinburgh. While not attending lectures, he had plenty of time to pursue his interests in collecting and natural history. He joined a number of student clubs and presented his first paper on a natural history subject to one of them, finding this type of study, learning on his own in a informal atmosphere, much more instructive than attending lectures. Through this route, he came to the attention of Dr Robert Grant, who had wide ranging interests across many scientific fields and became the first of a number of academic mentors to Darwin. In his autobiographical writing, Darwin would say of Grant:

> I knew him well; he was dry and formal in manner, but with more enthusiasm beneath this outer crust. He one day, when we were walking together burst forth in high admiration of Lamarck and his views of evolution. I listened in silent astonishment, and as far as I can judge, without any effect on my mind. I had previously read the *Zoonomia* of my grandfather, in which similar views are maintained, but without producing any effect on me. Nevertheless it is possible that the hearing rather early in life such views maintained and praised may have favoured my upholding them under a different form in my *Origin of Species*.

In the summer months Darwin continued to occupy himself with country sports, spending much of his time at

Maer, his uncle Jos's estate, and visiting neighbouring estates for the shooting. The social aspects of the sporting life also appealed greatly and there were always interesting and entertaining people to dine with in the evening. Emma Wedgwood would, no doubt, have been part of this company but, at this stage, there was no indication of any romantic feelings between the future husband and wife. Darwin was actually attracted to another estate owner's daughter called Fanny Owen. The two saw one another regularly while Darwin was in the East Midlands and wrote to each other when he was at college, although nothing would ultimately come of it. Shortly after Darwin embarked on the *Beagle*, Fanny would marry someone else.

Towards the end of the summer of 1827 some difficult decisions had to be made. It was obvious that Darwin was not suited to a medical career and his father insisted he find another suitable profession, worried that his feckless son would waste his life indulging in his favourite pastimes. With the legal profession and the military equally as unsuitable as medicine, Robert Darwin suggested what now appears the most unlikely of career choices. His son should become a clergyman.

Surprising as it may seem, the eighteen-year-old Darwin was himself quite enthusiastic about the idea of becoming a country vicar, even though he had never given the impression of being particularly religious. His decision to take religious orders had more to do with the social aspects of the Anglican clergy at the time than with religion. Country vicars were very much involved in the social life of their parishes and, as well as having a good and steady income, had plenty of free time on their hands to indulge their interests, such as country sports and natural history. Gilbert White had followed this path in the mid-eighteenth century and his book, *The Natural History of Selborne*, was not only well on its way to becoming an enduring classic of nature writing but was also one of the young Darwin's favourite books. It was probably not difficult for him to imagine himself living the life of a country vicar when he took people like Gilbert White as his models.

With this in mind, and after spending some months with a private tutor relearning the Latin and Greek he had forgotten while in Edinburgh, he went up to Christ's College, Cambridge at the start of the following year to begin a three-year Bachelor of Arts degree with the intention of going on to take holy orders afterwards. He found Cambridge an easy place to live socially, with several of his friends from Shrewsbury School already there, as well as his cousin William Darwin Fox, who would introduce Darwin to the joys of collecting beetles and become a lifelong friend. Initially, he fell in with the country sports set, spending more of his time shooting, drinking and dining than studying, but gradually he began to apply himself, particularly in the areas he had always found the most interesting – the natural sciences.

One of the standard texts in natural history at the time was *Natural Theology* by William Paley, first published in 1802. Darwin read this, together with other works by Paley, saying later that these were some of the few books of the ones he had to read for his degree that left any lasting impression on him. Paley advanced the argument of the divine nature of creation and wrote that the study of the natural world could lead to a better understanding of God. He used what would become a very well known analogy to illustrate the argument for design. He suggested that, if someone out walking were to find a watch without knowing how it had been made, they would reach the inevitable conclusion that the intricate mechanism had not come together by chance but had been designed and put together by a watchmaker. Organisms in the natural world, Paley argued, were a great deal more complicated than a watch, implying an intelligence at work in their design and thereby proving the existence of God. As an undergraduate Darwin did not question this reasoning and, even though he would contradict this argument in *The Origin of Species*, his own writing was influenced by Paley. He may also have first been introduced to the population theory put forward by Thomas Malthus through reading Paley. Malthus's ideas were an important influence on his later work and would be instrumental in leading him to his own theory of natural selection.

With the exception of Paley, Darwin remained sceptical about the use of much of the formal study he was required to do for his degree, preferring instead to find his own methods of learning: reading on his own account, making his natural history collections and entering into discussions with his peers. Gradually he moved away from the idle life of the sportsman he had dropped into at

Cambridge, becoming more serious and studious. This new attitude brought him to the attention of a number of his tutors, including the geologist Adam Sedgwick and the botanist John Henslow, both of whom were not only university professors but also ordained in the Anglican Church.

Darwin would remain far too modest throughout his life to articulate the reasons why such eminent scientists would take an interest in him, preferring instead to highlight his various failings as a student. However, he was bright and enthusiastic and, in his own studies, showed enough independence of mind to follow his own interests. He was also affable and entertaining company and was well used to mixing in the company of high society from his shooting days on the estates of Staffordshire. Henslow, in particular, thought highly of Darwin, perhaps seeing much of himself as a young man in a student who was beginning to make his way in science. The Friday evening get-togethers held by Henslow were something of a Cambridge institution and, after being introduced to them by William Darwin Fox, Darwin became a regular attendee. He also became a regular at Henslow's botany lectures, breaking the habit of not bothering to attend any of the formal studies of the university he had got into when he first arrived in Cambridge. He later wrote:

> Before long I became well acquainted with Henslow, and during the later half of my time at Cambridge took long walks with him on most days; so that I was called by some of

> the dons "the man who walks with Henslow"; and in the evening I was very often asked to join his family dinner.

By this time Darwin had become obsessed with collecting beetles, spending large amounts of time riding out to likely sounding locations to hunt for more specimens, sometimes employing assistants to carry his equipment and regularly writing to natural history magazines about some of his more unusual finds. In the regency period, natural history was highly fashionable and collecting expeditions were a regular occupation for even the highest levels of society. The skills Darwin developed in this fashionable pursuit would be extremely useful techniques he would later go on to use when he became a serious naturalist.

Henslow had one of his most important influences on Darwin by encouraging him to broaden his study of the natural world to include such disciplines as geology. He also suggested books to further Darwin's reading, including *A Preliminary Discourse on the Study of Natural Philosophy* by John Herschel, a book which emphasised the rational basis of science and the use of observation and experimentation in scientific research. Darwin would later describe it as a book that 'stirred up in me a burning zeal to add even the most humble contribution to the noble structure of science' and he would allude to Herschel in the opening paragraph of his introduction to the *Origin*, describing him as 'one of our greatest philosophers'.

Another book Darwin read at this time which had an equally lasting influence on him was *Personal Narrative of a*

Journey to the Equinoctial Regions of the New Continent by the great German scientist and explorer Alexander von Humboldt. The book, generally referred to simply as *Personal Narrative*, describes Humboldt's journey through the Americas between 1799 and 1804. His account contains details of, amongst many other scientific subjects, the geography, geology and natural history of the areas he visited, all of which came from his own personal observation. Darwin described him as 'the greatest scientific traveller who ever lived' and the influence of Humboldt's book on Darwin's account of his own journey are very apparent. The structure, style and content of *Journal of Researches*, or *The Voyage of the Beagle* as it would later become known, all owe a debt to Humboldt, although when Darwin first met the great man many years later he professed to being rather disappointed by him.

The first part of *Personal Narrative* concerns the extended stay Humboldt made on Tenerife, one of the Canary Islands. His vivid descriptions of the unusual geology and natural history of this island fired up Darwin's imagination and, as he came to the end of his three-year degree course in Cambridge, he began to make plans to mount an expedition to the island himself. Henslow advised Darwin to study the technical aspects of geological investigation before setting out and persuaded Adam Sedgwick to take him along on his annual summer excursion to study the geology of North Wales. They set out after Darwin had taken and passed his final exams, staying with Darwin's family in Shrewsbury on the way. The trip would only last a matter of ten days or so but, in that time, Sedgwick taught Darwin the basics of field geology which would prove invaluable to him in the future. It also made him realise that geology, rather than being a subject to be avoided, as he had mostly done since the dull lectures in the subject he had attended in Edinburgh, was an important means of studying the natural world.

Darwin would write in an autobiographical sketch:

> At Capel Curig [a village in what is now the Snowdonia National Park] I left Sedgwick and went in a straight line by compass and map across the mountains to Barmouth [on the coast of West Wales], never following any track unless it coincided with my course. I thus came on some strange and wild places and enjoyed much this manner of travelling. I visited some Cambridge friends who were reading there, and thence returned to Shrewsbury and to Maer for shooting; for at this time I should have thought myself mad to give up the first days of partridge shooting for geology or any other science.

As well as giving a basic grounding in geology, Sedgwick had also given Darwin the confidence to investigate geology and the natural world on his own and to formulate and develop his ideas rather than to rely completely on the descriptions of others. He returned to Shrewsbury keen to arrange the expedition to Tenerife, still planning to go back to Cambridge afterwards to begin the theological studies necessary to become a country vicar. When he arrived at Mount House, a letter from Henslow was waiting for him which would change his life and set him on the course that would lead him to becoming the most famous naturalist in the world.

A Five-Year Mission

Darwin opened the letter from Henslow on 30 August 1831. It contained details of an opportunity Captain Robert FitzRoy was offering for what he described as a 'gentleman companion' to accompany him as an unofficial naturalist on a surveying voyage to South America, which would also include a circumnavigation of the globe and was expected to last for a minimum of two years. The main purpose of the voyage would be to survey the coast of the southern part of the South American continent but, like all such naval surveying missions, there were numerous other secondary experiments and observations to be made, including, in this case, the testing of naval navigation techniques based on the use of chronometers.

Almost the entire continent of South America had become independent from Spain and Portugal in the early nineteenth century, culminating in Brazilian independence in 1825. Up until then the Spanish and Portuguese had monopolised trade with the continent, but, with independence, the region was beginning to open up to international commerce. Detailed navigational charts were an enormous advantage to those British interests looking to exploit these new opportunities, as was getting there before any commercial rivals from France or the USA. A further aim of surveying missions like that of the *Beagle* was to gather as much general information as possible: on the political climate, military capabilities, availability of natural resources and just about anything else that could be potentially useful in making decisions about future commercial or military activity in the region. The presence of a Royal Navy vessel in the region also sent out signals to any potential rivals. The *Beagle* was fitted out as a surveying ship but was, nevertheless, armed with cannon and carried a contingent of marines.

As part of the mission was to collect as much information as possible, a naturalist was included in the ship's company. The duties of naturalist were usually undertaken by the ship's surgeon and, initially at least, FitzRoy was really looking for a companion for himself more than anything else. Even so, Darwin immediately made up his mind he wanted to take up the position, even though it entailed paying his own way. In Darwin's case, this actually meant it would be at his father's expense. Robert Darwin, probably envisaging his son throwing away a respectable career in the clergy, initially refused to allow Darwin to go, although he did not rule the idea out entirely, telling him that he would reconsider if Darwin could find a man of common sense who would support the scheme.

The inference was clear. Robert Darwin would listen to the opinion of his brother-in-law, Darwin's Uncle Jos, whom he described as the most sensible man he had ever met. Darwin had already made plans to go to Maer for the start of the partridge shooting season on 1 September and travelled up to his uncle's estate the following day. With so much on his mind, he recorded in his diary, he could hardly shoot straight at all, only bagging one partridge, but Uncle Jos listened to Darwin's arguments in favour of taking up the offer and agreed with his nephew that it was far too good an opportunity to miss. Robert Darwin had made a list of all his objections and Josiah Wedgwood went through them all in a letter, discounting them and finding reasons why Darwin should go. In the end, rather than send the letter, he drove the thirty miles to Shrewsbury in his carriage and spoke to Robert Darwin in person, convincing him to allow Darwin to go. Robert Darwin agreed and, in doing so, also agreed to pick up the bill for his son for the whole enterprise.

The next and final hurdle for Darwin to get over before he was accepted on the voyage was to be approved by FitzRoy himself and by the Admiralty. Their first meeting went well, although FitzRoy would later joke with Darwin that he had nearly rejected him because of the shape of his nose. FitzRoy was acquainted with physiognomy, the theory that it was possible to deduce people's characters from their facial appearance and, according to FitzRoy at least, Darwin's nose indicated he lacked the resolution for an arduous journey.

Whatever reservations FitzRoy may have had about Darwin, he put them aside and accepted him as a travelling companion and unofficial naturalist. Darwin travelled down to Devonport, where the *Beagle* was being refitted for the voyage, since it was due to sail almost immediately. The next few months were a frustrating time for him as one delay after another prevented the ship from getting under way. He began to feel unwell and concealed his symptoms, worried that they might jeopardise his place on board, although they were, perhaps, signs of the stress-related condition from which he would suffer intermittently throughout his later life. Eventually, with the ship finally ready and the tides right, the *Beagle* sailed out of the harbour on 27 December 1831. The voyage was originally intended to last two years, but the disappearing coastline of Devon would be the last Darwin would see of England for almost five years.

Despite embarking on a circumnavigation of the globe, the *Beagle* was by no means a large vessel. She was a three-masted square-rigger of the class known by the navy as a brig sloop and was 90 feet long with a displacement of 235 tons. Even by the standards of the Royal Navy at the time, conditions would be cramped. The total number of people aboard as the ship left Devonport was 74, with 65 of these being the crew and the remaining nine being described by FitzRoy as 'supernumeraries', by which he meant people who were not in the naval service. As well as Darwin, the ship carried his servant, a steward for FitzRoy, a draughtsman to draw up maps from the survey and an instrument maker to look after the chronometers, of which there were 22 on board, and the other scientific equipment. Finally the ship also carried a missionary and three Tierra del Fuegians, two men and a young woman[12], who had been brought to England by FitzRoy at the end of his previous voyage to the tip of South America in order to be 'civilised' with a Christian education. They were now being returned to Tierra del Fuego to help in FitzRoy's plan to establish a mission there.

FitzRoy had been appointed to the captaincy three

years before after the previous captain had committed suicide while the ship had been surveying the coastline of Tierra del Fuego. The reasons for the suicide had not been totally clear, but they appeared to have been the result of severe depression brought on by the extremely difficult conditions in the South Atlantic, in conjunction with loneliness and isolation, common problems for Royal Navy captains on long voyages. This was one of the reasons why FitzRoy decided to take a companion with him. As the nephew of the high ranking Anglo-Irish politician and diplomat Viscount Castlereagh, whose mental breakdown in 1822 had led to his suicide, he was also well aware of a history of mental instability in his own family. FitzRoy was said to bear a close physical resemblance to his deceased uncle and was concerned that he might also share some of his psychological traits.

In an era when patronage was one of the recognised ways of advancing a career in the navy, FitzRoy's aristocratic background had stood him in good stead. He also had the support of Francis Beaufort, the head of the Hydrographic Office of the British Admiralty and another Anglo-Irish aristocrat. Some in the Admiralty had thought FitzRoy's decision to bring four Fuegians back to England during his last voyage had displayed a lack of sound judgement but Beaufort supported FitzRoy and approved the new voyage, including Darwin's participation in it. He also wanted FitzRoy to test the scale he was developing to measure wind speed, which he aimed to introduce as the standard measure throughout the navy. The Beaufort Scale would later become the standard across the Royal Navy and, having been modified on a number of occasions, is now used around the world.

Without the patronage of such people as Francis Beaufort, FitzRoy could not possibly have advanced his career as quickly as he did. He was only 23 years old when he assumed the captaincy of the *Beagle*, an extraordinarily young age for such a responsibility, and was 26 when he was commissioned to return to South America. But he was also a highly capable and intelligent naval officer, one of a new generation who had attended the Royal Naval College in Portsmouth, where he had passed out with distinction. He was a man liable to extreme outbursts of temper, and officers who served under him developed a code between themselves to assess his mood which involved an enquiry as to whether the captain had drunk a cup of coffee in the morning, when they thought his moods were at their worst. However, he could also be charming and, as Darwin would describe him, 'generous to a fault'.

Darwin experienced FitzRoy's temper first hand on a number of occasions during the voyage, most notably after getting into an argument with him about slavery. The slave trade was abolished in the British Empire in 1807 but, as a measure to ensure slave owners did not lose money, the actual practice of slavery itself was not abolished until 1833 and continued long after that in some countries which were not part of the empire. Darwin, following both his father and grandfather, was a supporter of the abolitionist Whig party, while FitzRoy, also following the family tradition, was a high Tory, who was very much against abolition. Darwin described the argument:

> ... early in the voyage at Bahia [now Salvador] in Brazil he defended and praised slavery, which I abominated, and told me that he had just visited a great slave-owner, who had called up many of his slaves and asked them whether they were happy, and whether they wanted to be free, and all answered "No". I then asked him, perhaps with a sneer, whether he thought that the answers of slaves in the presence of their master was worth anything. This made him excessively angry, and he said that as I doubted his word, we could not live any longer together.

Darwin thought he would have to leave the ship as a result of the argument but, with what he describes as FitzRoy's 'usual magnanimity', the captain apologised and asked Darwin to continue living with him.

The occasional argument aside, Darwin and FitzRoy remained on good terms for the entire voyage. The same could not be said for Darwin's relationship with Robert McCormick, the ship's surgeon, who considered himself to be the official naturalist on the voyage and quickly grew to resent what he saw as the preferential treatment afforded to Darwin by FitzRoy. On a number of occasions when the ship arrived at an island, FitzRoy would invite Darwin to accompany him ashore while leaving McCormick on board and he also allowed Darwin to send some of his collection back to England at the Admiralty's expense, even though Darwin had no official position on the *Beagle* and his collection, which was worth a large amount of money, would remain his own property. (Darwin had no intention of selling his specimens, as he was intending to use the collection to promote his own scientific career when he returned to England, but he had expected to pay all his own expenses, including those for the carriage of the collection, and FitzRoy's offer was an unexpected and very generous bonus.)

It is, perhaps, understandable that McCormick was annoyed. He would have recognised the opportunity to further his own career that came with making a natural history collection on such a voyage and he must have felt his position had been usurped by a young Cambridge toff who wasn't even in the navy. Whatever McCormick thought, Darwin remained unconcerned and was quite disparaging about the surgeon's abilities as a naturalist in the letters he wrote to his family. In truth, McCormick does not appear to have shown anything like the vigour and enthusiasm for natural history and for collecting specimens that Darwin did and, after only four months on board, he decided to leave the ship in a fit of pique and return to England.

From then on, Darwin was unchallenged in his position as naturalist on the ship and became known as 'Philos', short for Ship's Philosopher, to the officers on board. His enthusiasm for the subject rubbed off on a number of the crew, including the captain, who began their own collections. In one or two areas, these collections actually ended up better than his own. After Darwin had returned to England and it had been pointed out to him that each island in the Galapagos archipelago appeared to be inhabited by a different species of finch, for example, he had to borrow FitzRoy's collection to confirm that this was the case because he had not labelled his own carefully enough to know for sure.

This is not to say Darwin often made these sort of mistakes. His experiences as a country sportsman and a beetle collector, combined with the time he had spent with John Henslow and Adam Sedgwick in Cambridge, proved an almost perfect apprenticeship for his work in collecting specimens and he would become a very fine naturalist. And, wherever he went, he had a copy of Humboldt's *Personal Narrative* with him, which he used almost as if it were a guidebook.

If he thought of himself as travelling in his hero Humboldt's footsteps, Darwin faced initial disappointment. The first port of call for the *Beagle* was to have been Tenerife, the island he had been planning to visit after reading about it in his copy of Humboldt's book. But, when the *Beagle* reached the island, nobody was allowed to go ashore before a twelve-day period of quarantine had been served, because of the threat of a cholera epidemic spreading to the island. FitzRoy was not prepared to wait and sailed on to the Cape Verde Islands.

Darwin had another reason for wanting to get back onto dry land as soon as possible. From the moment the *Beagle* had set sail he had suffered terribly with seasickness, finding the only way he could get any relief was by lying as motionless as possible in his hammock. For the entire five years of the voyage he never really gained any sea legs. Seasickness was an occupational hazard for the whole crew, but FitzRoy was particularly sympathetic to Darwin's plight. Whenever possible he put his companion on shore, making arrangements to pick him up when the *Beagle* was moving on and keeping him informed about the ship's schedule so he could make his way to the next port of call overland. In the course of the voyage, Darwin spent less than a third of his time actually on board ship.

After more than two weeks continuous sailing Darwin must have been desperate for the *Beagle* to arrive at St Jago, now known as Santiago, in the Cape Verde Islands, an archipelago of volcanic outcrops about 400 miles off the coast of West Africa. Some of the officers and crew had been to St Jago before and told Darwin it was mostly made up of barren and sterile volcanic rock, but Darwin was entranced by the tropical vegetation he found there. The island had been settled by the Portuguese for 300 years by that time and the domestic animals they had introduced had severely depleted the natural vegetation, but his first experience of an oceanic island affected him profoundly. After the first day he spent on St Jago, he wrote in his diary 'It has been for me a glorious day, like giving eyes to a blind man'.

Although in later years Darwin's name would become much more associated with the Galapagos Islands, he himself would always remember the visit to St Jago as a turning point in his life. While on the island, he decided he was going to write a book on the subject of the geology of volcanic islands, which he would actually do when he had returned to England, and it is possible to see this as the moment when he left behind any previous thoughts he had about the direction his life would follow and dedicated himself to science.

Before leaving England, FitzRoy, who shared Darwin's interested in geology, had given his companion a copy of *Principles of Geology* by Charles Lyell. The book had been published in the previous year and set out Lyell's views on geological science which, by the 1830s, had become a much more scientifically based discipline than natural history, then some distance away from developing into the biological sciences. Little of what Lyell set out was actually new, being based on the work of late eighteenth century geologists such as James Hutton, now regarded as the father of modern geology. Lyell's achievement was to bring these ideas together and make them understandable. In essence, he described the surface of the earth as being made up of a crust which was subject to gradual shifts, raising and lowering geological features over very long periods of time. This theory was known as Uniformitarianism, the central idea being that the forces acting on the Earth's crust were natural processes which had occurred over many millions of years and were continuing to occur. This contrasted with a rival theory, Catastrophism, developed by the French naturalist George Cuvier, which held that the Earth had been created by God and had been shaped by sudden enormous events and forces which were no longer occurring.

The influence Uniformitarianism would have on Darwin's theory of natural selection is immediately apparent. It too hypothesised change by gradual and ongoing natural forces rather than by a sudden creation, although

Darwin and Lyell did not totally agree with each other even when they became close friends after Darwin returned to England. As well as such gradual processes as uplift and erosion, Darwin thought that rapid change could also occur through natural events such as volcanoes and earthquakes, both of which he would experience directly later in the voyage. But on St Jago, armed with Lyell's *Principles of Geology*, Darwin saw geological features in terms of gradual change. The theoretical framework given to him by Lyell was enhanced by the practical knowledge he gained from surveying techniques used by the *Beagle* crew. On board the ship he shared a cabin with two of the young surveying officers, John Lort Stokes, who was nineteen, and Philip Gidley King, only fourteen when the *Beagle* left England. The cabin was small and the table Stokes used to draw up charts took up most of the space. Darwin could hardly fail to pick up some knowledge of the methods of surveying and, with Stokes keen to show him the techniques involved, Darwin learned to look at the landscape with a practical eye as well as a theoretical one.

The independence of mind Darwin had shown at Cambridge was another factor in the way he began to look at the geology of the landscape of St Jago. Although guided by Lyell, he did not slavishly follow everything the older man wrote. Instead he examined the geology in front of him and was prepared to come to his own conclusions even if they were in conflict with Lyell's views. This ability to rely on himself, on his own observations and experiments, would be one of the key reasons why Darwin would be able to develop theories which were not only new but flew in the face of received ideas.

While he was on St Jago, Darwin also began his natural history collections, building on his experience collecting beetles in Cambridge. The methods he employed were essentially the same as before, including using an assistant to do much of the leg work. At first his servant, Henry Fuller, who was one of FitzRoy's cabin staff, fulfilled the role, but Darwin would later employ a member of the crew, the seventeen-year-old Syms Covington, whom he paid out of his own pocket, as a full time assistant. Covington would work for Darwin throughout the remainder of the voyage and for several years after they had returned to England, although he would hardly get a mention in Darwin's account of the voyage. The relationship between them was very much one of gentleman and servant. Even though they spent long periods of time in each other's company and Covington accompanied Darwin on most of the overland expeditions he undertook on the South American mainland, they appear to have had very little to do with each other on a personal level.

From St Jago, the *Beagle* headed south west, crossing the equator, where both FitzRoy and Darwin had to undergo the traditional navy 'Crossing the Line' ceremony, in which they had buckets of water thrown over them and Darwin, since it was his first time, had his face painted and was shaved by members of the crew. They arrived in Bahia on the coast of northern Brazil at the end of February, where Darwin had his first experience of a tropical forest, and then the *Beagle* headed south to Rio de Janeiro.

FitzRoy was engaged in checking Admiralty charts for the Brazilian coastline, sorting out some discrepancies that had arisen from previous surveys, and Darwin spent almost the whole period on shore. Along with a number of other members of the *Beagle* company, he rented a cottage at Botofogo Bay, then a few miles outside Rio de Janeiro. One of his companions was Augustus Earle, the ship's artist, who had visited the city previously and introduced Darwin to the sights. As Darwin would find almost everywhere he went, association with a captain in the Royal Navy opened many doors to him that would have remained closed to an independent traveller. He immediately became part of the expatriate social scene, and the British contingent in the city proved only too happy to help him in any way it could. But, despite the introductions and invitations he was constantly receiving, what Darwin was most interested in doing was continuing his geological research and expanding his natural history collections, which now included fossils as well as the specimens he had caught or shot. Accordingly he used the cottage as a base from which to travel along the coast and into the interior of Brazil.

By July, FitzRoy had completed the work he had been assigned in Brazil and Darwin rejoined the *Beagle* as it headed further south, along the coast of Uruguay and Argentina. Much to FitzRoy's annoyance, the *Beagle* was fired on by an Argentine ship as it approached Buenos Aires, although Darwin reacted as if it was all part of the adventure. The shot had been a blank, but FitzRoy was not about to let anyone get away with shooting at a Royal Navy ship. On arrival at Montevideo, on the other side of the River Plate from Buenos Aires, he reported the matter to the captain of a British frigate which was stationed there, who immediately set sail to set the Argentinians right on the matter. FitzRoy appears to have been in a bullish mood while in Montevideo because, when a request came from the Uruguayans to help put down a minor army insurrection, he armed and dispatched almost the entire crew to deal with the situation. Darwin went along to observe the proceedings but, much to his apparent disappointment, the insurrection ended without a shot being fired.

The coastline to the south of Buenos Aires was less well known than to the north and, on leaving Montevideo, FitzRoy could begin the main business of the voyage. As usual, Darwin spent most of his time on shore, even though the Argentinian pampas, the huge area of grassland in the interior of the country, was relatively lawless at that time. He went on long trips on horseback, accompanied by other members of the crew and sometimes by the local gauchos, and either stayed at one of the huge *estancias*, as the cattle ranches of the region were called, or camped out on the pampas. These trips could go on for weeks. On one of them, for example, he travelled from Bahia Blanca to Buenos Aires, a distance of about 500 miles.

The Argentinian army, under the command of General Juan Manuel de Rosas, was engaged at the time in a campaign against the indigenous peoples of the pampas, whom Darwin described as the 'wild Indians', the purpose of which was their complete extermination. On one of his trips Darwin encountered the general, who would subsequently become, in effect, the dictator of Argentina, and was highly impressed by him, describing Rosas as the most prominent man in South America.

As well as collecting specimens, Darwin was often asked to accompany members of the crew on hunting expeditions, attempting to bag some game to supplement what could become a monotonous diet on board ship. Darwin's shooting skills, honed on his uncle's estate in the partridge season, came in very useful for this purpose and he also became more familiar with the habits of the animals he hunted, including pampas deer, guanaco (a relative of the camel similar to the llama), armadillos, which Darwin thought were delicious, and the rhea, a large flightless bird not unlike the ostrich. He was particularly interested in the differences between the two species of rhea in South America which shared most characteristics in common but, despite an overlap in geographical distribution, tended to occupy slightly different regions. Darwin had been alerted to the difference between the two types of rhea by gauchos (who also told him of the females' habit of laying eggs in a communal nest which was then solely guarded by the male) but it took several months of looking before he came across the smaller of the two. This species is now known as Darwin's rhea and he finally recognised one during an evening meal at a camp site while he was in the process of eating it, the bird having been shot during the day without him realising what it was. He immediately collected all the parts that had not been eaten, including the skin, feathers and the bones, and put them back together to form a skeleton as well as he could, before adding it to his collection. Darwin would later make use of the example of the rhea – and the fact that two separate species could co-exist in the same place while remaining distinct – in *The Origin of Species*.

By December 1832 FitzRoy was ready to sail further south, making for the series of islands off the southern tip of the South American continent known as Tierra del Fuego during the summer months when the otherwise notoriously unpredictable and difficult weather conditions would be at their best. At that time there were no European settlements at all in the extremity of the continent and, for the first time, Darwin met some of the indigenous inhabitants of the continent. In his journal he recorded their way of living and, at different times, was amazed and appalled by them, describing them, using the language of the period, as 'savages' and their way of life as 'primitive' while acknowledging the difficulty of surviving at all in such an inhospitable environment.

Tierra del Fuego was largely unexplored by Europeans and, in recording the geology and natural history, Darwin felt like he was breaking new ground in the tradition of some of the great explorers of the past, such as Ferdinand Magellan and Captain James Cook. FitzRoy continued with his surveying mission, mapping the uncharted coastline, but he was particularly concerned with his own personal attempt to establish a mission to bring Christianity to what he saw as a godforsaken and blighted land. Accordingly he found a suitable site on a relatively sheltered inlet and directed the crew to build a wooden mission house. Robert Matthews, the missionary FitzRoy had brought from England, together with the three Fuegians, was left at the mission while the surveying continued but, on returning a few weeks later, FitzRoy found that Matthews had been attacked by the local inhabitants and his possessions had all been stolen. He was in fear of his life and desperate to be taken back onto the *Beagle*. The following year the *Beagle* returned again to find that the mission was completely destroyed and FitzRoy's 'civilised' Fuegians had resumed their original way of living. FitzRoy realised his attempt to bring Christianity to the region had been a complete failure and, rather than continue any further, he decided regretfully to give up his plans.

The *Beagle* did not stay in the vicinity of Tierra del Fuego permanently but moved up and down the coast, returning as far as Buenos Aires in the winter months. FitzRoy also made two trips to the Falkland Islands, about 280 miles off the coast of southern Argentina. Neither he nor Darwin were particularly impressed by the Falklands, finding them desolate and windblown, and their mission there had as much to do with politics as it did with the small amount of surveying FitzRoy had to do. The sovereignty of the Falklands was as contentious then as it remains now and the presence of a British naval ship, even if it was not a warship but only a surveying vessel, sent signals to the Argentinians that Britain intended to hold on to its possession.

As always, Darwin spent much of his time on shore examining the geology of the islands and collecting specimens. He recorded his encounters with the warrah, a canine animal he describes as being somewhere between the size of a fox and a wolf. He noted the difference between the warrahs on East Falkland and those on West Falkland, suggesting it was another example of divergence and another reason to study the natural history of remote islands. He also thought the warrah would go the way of the dodo because it threatened the livestock of the Falkland Island settlers. His prediction did not take long to become a reality, and the warrah had been hunted to extinction by 1876.

Writing in his journal about the time he spent in the Falklands, Darwin would include a long passage on the kelp forests surrounding the islands. It is significant because of the way Darwin described the various inhabitants of this environment as making up a community – what today would be called an ecosystem. He compared these kelp forests, along with the ones he examined around Tierra del Fuego, to the tropical forests he had seen in Brazil and went on to write:

> One single plant forms an immense and most interesting menagerie. If this Fucus [the kelp] was to cease living, with it would go many of the seals, cormorants and certainly the small fish and sooner or later the Fuegan Man [Darwin's name for the indigenous people of Tierra del Fuego] must follow. The greater number of invertebrates would likewise perish, but how many is hard to conjecture.

Rather than simply recording the species he had seen and collected in the kelp, Darwin recognised the diversity of life in this environment and the importance of the interactions between the different species. As well as being an example of 'the struggle for life', the term Darwin used to describe the competitive nature of existence which he considered to be the driving force of natural selection, it was a very modern way of describing the natural world.

By June 1834, FitzRoy had finished all the surveying he had to do along the Atlantic coast of South America and the Falklands. The *Beagle* navigated its way through the Strait of Magellan, the sound which separates Tierra del Fuego from the mainland and which offered a less hazardous passage during the winter months than going round Cape Horn, and entered into the Southern Pacific Ocean. The ship sailed north to Valparaiso, a cosmopolitan city and trading port on the Chilean coast and, after the rough weather he had experienced in Tierra del Fuego, Darwin was delighted to be ashore in a region with a much more comfortable Mediterranean climate. FitzRoy continued with the survey. While in Valparaiso, Darwin stayed with Richard Corfield, whom he had known at Shrewsbury School, proving that, even in the 1830s, the world could be a surprisingly small place. Corfield was happy to help Darwin in any way he could, going on trips into the Andes with him and introducing him to other people who could be useful to him.

During his time in Valparaiso an incident occurred which very nearly ended Darwin's participation in the voyage before he had the opportunity to go to the

Galapagos Islands. The previous year FitzRoy, without checking with the Admiralty first, had bought a schooner with his own money to help with the surveying work. When his actions came to light, the Admiralty refused to pay for the schooner, an action that cast FitzRoy's judgement into question again. He became increasingly depressed at this turn of events and decided to resign his command and leave the ship. Since Darwin was his guest, this would have meant Darwin going with him. FitzRoy's second in command, Lieutenant John Clements Wickham, talked FitzRoy out of doing anything impetuous even though he stood to take command of the vessel himself.

After returning to Corfield's house in Valparaiso from one of his many trips into the Andes, Darwin succumbed to an illness which lasted for a month and required him to spend most of that time in his bed. It was the first occasion such an illness had incapacitated him. He put it down to the after effects of drinking a bad bottle of wine, although it could also have been the beginning of the problems he would experience with ill health throughout his later life.

A trip southwards on board the *Beagle* early in 1835 appeared to do both Darwin and FitzRoy good, helping them to get over their various physical and mental problems. They sailed to the island of Chiloé and, from a safe position on the island, watched the successive spectacular eruptions of two volcanoes on the mainland. A short time later they had crossed back to the mainland and were in the small coastal town of Valdiva when an earthquake struck. The centre of the quake was near the city of Concepción to the north but, even so, Darwin felt the earth rocking under his feet, making him aware of the tremendous power involved in these natural geological processes. They travelled up to Concepción to find that many buildings had been destroyed and more than 100 people killed.

Both Darwin and FitzRoy used the opportunity to study the geology of the area where the earthquake and volcanoes had struck in the context of Lyell's theories as set out in the *Principles of Geology*. FitzRoy's account of the events were read out at a meeting of the Royal Geographical Society, of which Lyell was the president, and Darwin would write extensively about his theories of the uplift caused by these ongoing geological events, along with the sedimentation and erosion that had formed the spectacular landscape of the Andes. In doing so, he was demonstrating the confidence of an experienced field geologist, using Lyell as a guide but coming to his own conclusions about the formation of these geological features.

The *Beagle* had been away from England for more than four years by this time and Darwin considered returning home by crossing the South American continent to Buenos Aires and returning to England by ship. FitzRoy had finally finished all of the surveying he needed to do along the Pacific coast and was preparing to embark on the circumnavigation of the globe which would culminate, of course, with the ship arriving back in England. Darwin changed his mind and stayed on board as the *Beagle* headed along the coast of northern Chile and Peru and then headed eastward out into the Pacific.

The first stop on the voyage home was at the Galapagos Islands, a group of volcanic islands straddling the equator about 600 miles off the mainland of South America. Of all the places Darwin visited during the voyage, the Galapagos have become most associated with his name. His experiences on the islands have been recounted numerous times and some of the accounts have included a version of events not really borne out by his own writings from that time. To suggest that Darwin's theories concerning natural selection and the struggle for life came to him in a flash of inspiration while he was in the Galapagos is an invention. Such stories about breakthroughs in science are common, making for a more exciting and easily explained scenario than the painstaking process of developing an idea over a long period of time and then testing it thoroughly with observation and experimentation.

In later life Darwin would do little to dispel this simplified version of events, in which he had a 'eureka' moment on the Galapagos, and there can be no doubt that the time he spent there would be important to him as he developed his theory of natural selection, but the journal he kept at the time and the notebooks he would keep after he returned to England show that he did not arrive at his theory until several years after his experiences on the islands. In the introduction to *The Origin of Species* he actually makes this clear:

> When on board HMS *Beagle*, as naturalist, I was much struck with certain facts in the distribution of the inhabitants of

> South America, and in geological relations of the present to the past inhabitants of that continent. These facts seemed to me to throw some light on the origin of species – that mystery of mysteries, as it has been called by one of our greatest philosophers. On my return home, it occurred to me, in 1837, that something might perhaps be made out of this question by patiently accumulating and reflecting on all sorts of facts which would have any bearing on it.

It is hard to know why anybody would ignore such a prominent passage in Darwin's major work to create the story that natural selection came to him like a bolt from the blue. In truth, Darwin spent only 36 days in the Galapagos in total and, on 17 of those days, he was on board the *Beagle* as it sailed between the different islands and carried out surveying work. This is not to say that he was not struck by the remarkable geology and natural history he encountered on the islands, although his initial reaction to their volcanic landscapes was not entirely positive. On 17 September 1835, two days after he arrived, he wrote in his journal of Chatham Island, now known as San Cristóbal:

> Nothing can be less initially inviting than the first appearance. A broken field of black basaltic lava is everywhere covered by a stunted brushwood, which shows little sign of life. The dry and parched surface, having been heated by the noonday sun, gave the air a close and sultry feeling, like that from a stove: we fancied even the bushes smelt unpleasantly.

These initial impressions quickly gave way to a deeper appreciation, as he would make clear later in the entry in his journal on the same day:

> The natural history of this archipelago is very remarkable: it seems to be a little world within itself; the greater number of its inhabitants, both vegetable and animal, being found nowhere else.

Darwin also found the geology he encountered intriguing, the islands having been formed in relatively recent times by volcanic activity which, he observed, was very obviously still continuing. The climate was also unusual considering the position of the islands on the equator, since it was moderated by currents of cold water coming up from the southern Pacific Ocean. These unusual conditions, together with the remoteness of the islands from the mainland, are what has led to the unique array of life, with a high number of endemic species, both animals and plants, which are not found anywhere else on Earth.

Darwin particularly remarked on two features of the animal life on the islands. His job of catching specimens for his collection was made very easy because almost all the species he came across were incredibly tame, being unaccustomed to the presence of people. This was a characteristic he had noted in the animals he had encountered on other remote islands, particularly the Falklands. The other feature was the fact that, although the species he found on the islands were unique, they bore a resemblance to those he had already come across on the South American continent. After having some time to reflect on what he had seen, he wrote:

> I will not here attempt to come to any definite conclusions, as the species have not been accurately examined; but we may infer, that, with the exception of a few wanderers, the organic beings found on this archipelago are peculiar to it; and yet that their general form strongly partakes of an American character. It would be impossible for any one accustomed to the birds of Chile and La Plata to be placed on these islands, and not to feel convinced that he was, as far as the organic world was concerned, on American ground.

The full significance of the relationship of the animals on the Galapagos to those on the mainland would not become apparent to him until later. What he was describing is now known as speciation, which occurs when individuals of a species become isolated from the main population of that species and, under a different set of environmental conditions, evolve away from the original species.

Darwin would eventually realise not only that some groups of animals, including the various different species of giant tortoises, finches and mockingbirds, were related to the species on the mainland, but also that they did not occur over the whole archipelago. Different species were found on different islands. He was alerted to this phenomenon by Nicholas Lawson, the British Representative on the islands, who explained to him that it was possible to tell which island a tortoise had come from by the shapes of their shells and the patterns on them, which were different for the animals of each island.

The amount of time Darwin spent in the Galapagos was dictated, as it had been throughout the voyage, by

FitzRoy's surveying schedule. As soon as he was finished, the *Beagle* left the archipelago to make the long passage of more than 3,000 miles across the Pacific to the South Sea island of Tahiti. From there the ship sailed to New Zealand and then crossed the Tasman Sea to Sydney in Australia. By this time both FitzRoy and Darwin, and no doubt the rest of the crew as well, were keen to get back to England and they did not linger in any of their ports of call for longer than was absolutely necessary. The *Beagle* made brief stops in Tasmania and Western Australia before heading out into the Indian Ocean, sailing in a north westerly direction to the Keeling Islands, a collection of coral atolls and islets about half way between Australia and Sri Lanka which now go by the official name of the Cocos (Keeling) Islands. Darwin had already developed his own theory of how coral islands were formed before he had actually visited any examples and was delighted when the observations he made on the Keeling Islands confirmed what he had thought.

After ten days on the Keeling Islands, the Beagle sailed to Mauritius and then on to Cape Town, where Darwin dined with John Herschel, one of his scientific heroes, who was staying on the Cape to make astronomical observations of the skies in the southern hemisphere. The *Beagle* then rounded the Cape of Good Hope and, by way of St. Helena and Ascension Island, crossed the Atlantic again to arrive at Bahia in Brazil, the first port of call it had made in South America four and a half years previously. The ship stayed there for only four days before embarking on the final leg of the voyage, heading for the Azores in the North

Atlantic, before arriving in Falmouth on 2 October 1836. The *Beagle* – and Darwin – had been away from England for four years and ten months.

A Twenty-Year Wait

In the five years Darwin had been away on the *Beagle*, he had become a highly experienced naturalist and geologist, confident enough in his abilities to draw his own conclusions from what he had observed rather than simply relying on extending the theories of others. On the long ocean passages of the homeward leg of the journey he must have had plenty of time to contemplate the future, particularly what he would do when he returned to Britain. When the ship arrived at Ascension Island, a letter was waiting for him from his family telling him that Adam Sedgwick had stayed with them in Shrewsbury and told them that Darwin could become a great man of science. If he still entertained any thoughts of returning to his studies with a view to a career in the church, which seems highly unlikely, the opinions of Sedgwick convinced him to dedicate himself to science. He had known since his days as a medical student at Edinburgh that his father's wealth would allow him to follow any path he chose and the path opening up to him after leaving the *Beagle* for the last time was to follow in the footsteps of his scientific mentors and heroes, such as Humboldt, Herschel and, most particularly, Lyell.

After a short period with his family in Shrewsbury, Darwin spent three months in Cambridge unpacking and sorting out the huge collection of specimens John Henslow had been storing for him. Both Henslow and Sedgwick had been working on his behalf while he had been away, promoting his name in scientific circles and ensuring his activities as a geologist and naturalist were well known even before he returned. Darwin found himself something of a celebrity, much in demand at dinner parties and invited to attend meetings of scientific societies. Charles Lyell, whom Darwin met for the first time soon after he got back, was impressed with his work, not least because Darwin had taken up Lyell's ideas so avidly, and treated Darwin as his protégé. When Darwin moved to London, taking rooms in a house in Great Marlborough Street, within easy walking distance of both his brother and Lyell, he began to meet with Lyell on an almost daily basis. Lyell introduced him to London scientific society, putting him up for membership of, amongst others, the Geological Society of London, of which Lyell was the president. Darwin went on to read a number of papers at Geological Society meetings.

The next few years would be ones of feverish activity for Darwin. He distributed his collection to established experts to examine and describe, including the fossils to his future adversary Richard Owen and the birds to John Gould, a highly respected ornithologist from the Zoological Society of London. These descriptions, which included numerous species new to science, would form the basis of *The Zoology of the Voyage of H.M.S. Beagle*, a lavishly illustrated and handsome set of five volumes, published between 1838 and 1843 and edited by Darwin, a task which would take up much of his time. He was also preparing his journal for publication, initially as the third volume of FitzRoy's official account of the voyage, but subsequently as a book in its own right.

Darwin's journal would cause something of a rift between himself and FitzRoy, who was annoyed at what he considered to be Darwin's almost complete lack of appreciation and acknowledgement of all the help he had received from the captain and crew of the *Beagle*. They had already drifted apart after the voyage – FitzRoy had been married and the two men moved in entirely different social circles – and FitzRoy's displeasure over Darwin's journal would compound their differences, which would never be really repaired. In later years FitzRoy would become more religious, although he was never the zealot he is sometimes portrayed as becoming, and, despite an illustrious career, he was prone to frequent bouts of depression. An unhealthy obsession with the suicide of his uncle, Viscount Castlereagh, which was one of the motivating factors behind him taking Darwin onto the *Beagle* in the first place, would also continue throughout his life. It may even have prompted his own suicide in 1865, at the age of sixty, by cutting his throat with a razor, mirroring the method used by his uncle.

At the same time as Darwin was working hard on his various 'Beagle' projects and preparing papers on geology and natural history for learned societies, he made a decision that it was time he got married. He was approaching his thirtieth birthday and, after weighing up the advantages and disadvantages of married life, he felt the time was right to settle down and begin a family. The only problem was identifying his future wife. Charles Lyell was married to the eldest daughter of another well-respected geologist, Leonard Horner, who also had four younger and eligible daughters and, for a brief period, it looked as if Darwin would choose one of them. However, in November 1838, after a few visits to his Uncle Jos's estate in Maer and slightly out of the blue, he proposed to his cousin Emma Wedgwood. Apparently a little to his surprise, she accepted. They had known each other since childhood and, although there had been little indication of any great attraction between them and Darwin's proposal had an air of calculation about it, their marriage was, at least, built on a solid base of friendship. The marriage took place on 29 January 1839 at the Anglican church in Maer and, on returning to London, the newly married couple moved into a more suitable house in Gower Street for the two of them than the bachelor lodgings where Darwin had lived in Great Marlborough Street.

As if there was not enough going on in his life at the time, what with the ongoing work on the Beagle collection, the books he was preparing for publication and his forthcoming marriage, Darwin was also engaged in an intense period of private study, one he would keep almost entirely secret and which would lay the foundations of the work he would do for the rest of his life.

On board the *Beagle* Darwin had got into the habit of keeping a notebook as well as a journal, recording observations, thoughts, useful references in books and anything else that occurred to him. It was a habit he continued after the voyage was over, at first using a notebook he had begun while on board the ship, which he called the Red Notebook, and then in a series of seven other notebooks he labelled from A to D and N and M, which are now usually referred to as the Transmutation Notebooks. It is not entirely clear when he first began to think about the transmutation of species and he would later write that, while he was on the *Beagle*, his thoughts remained in line with the prevailing view that species were immutable or unchanging. It is impossible to say exactly when he changed his view but what is clear is that, by the middle of 1837, almost all the entries he was making into his notebooks were concerned in some way with transmutation.[13]

In a autobiographical sketch written forty years later, Darwin would recall:

> After my return to England it appeared to me that by following the example of Lyell in Geology, and by collecting all facts which bore in any way on the variation of animals and plants under domestication and nature, some light might perhaps be thrown on the whole subject. My first notebook was opened in July 1837. I worked on true Baconian principles, and without any theory collected facts on a wholesale scale, more especially with respect to domesticated productions, by printed enquiries, by conversations with skilful breeders and gardeners, and by extensive reading... I soon perceived that selection was the keystone of man's success in making useful races of animals and plants. But how selection could be applied to organisms living in a state of nature remained for some time a mystery to me.

When Darwin says he used the 'true Baconian principles', he is referring to the philosophy of Francis Bacon. In a nutshell, what he is saying is that he avoided any speculation, sticking instead to known facts and to what could be inferred from direct observation. Going against this purely rational method and in the absence of the facts, it is possible to speculate on how Darwin came to accept the concept of the transmutability of species. In March 1837, John Gould informed Darwin about the results of his examination of the finches and mockingbirds Darwin had collected in the Galapagos, telling him that they were each individual species. With the help of the specimens collected by FitzRoy and other members of the *Beagle* crew, Darwin could establish that each species came from a different island. He already appreciated that these species were similar to, but not the same as, those from continental South America, and he knew that the endemic species he had collected from St Jago in the Cape Verde Islands bore a similar relationship to other species from Africa, the nearest continental landmass to the island. The challenge Darwin faced was to find an explanation for these observable facts and the most straightforward conclusion he could reach was that the species on the islands had evolved away from a common ancestor which had arrived from the mainland. Modern evolutionary biology has shown this to be the correct interpretation but, at the time, it was a dangerous idea which bordered on heresy, so Darwin kept his conclusions mostly to himself, only discussing them with close and like-minded friends.

As has already been indicated, this is a speculative account of how Darwin may have come to accept the transmutability of species and it is, of course, entirely possible that he arrived at the idea by a totally different route. But, once he had accepted the concept, the obvious next step dictated by scientific principles was to find out how species changed or, to put it in more scientific language, to elucidate the mechanism of evolutionary change.

From July 1837, Darwin's notebooks are almost exclusively concerned with the problem of finding a mechanism of change. They are bursting with what might be taken for random facts and half-completed thoughts which, when considered in retrospect, show the direction in which his mind was moving. More than a year later, on 28 September 1838, Darwin wrote down his initial thoughts on reading *An Essay on the Principle of Population* by Thomas Malthus[14], which, as related in the previous chapter, would later be the spark he needed to bring his thoughts together to form a theory. In the autobiographical sketch from 1876 he wrote:

> In October 1838, that is, fifteen months after I had begun my systematic enquiry, I happened to read for amusement 'Malthus on *Population*', and being well prepared to appreciate the struggle for existence which everywhere goes on from long-continued observation of the habits of animals and plants, it at once struck me that under these circumstances favourable variations would tend to be preserved, and unfavourable ones destroyed. The result of this would be the formation of new species. Here, then, I had at last got a

> theory by which to work; but I was so anxious to avoid prejudice, that I determined not for some time to write the briefest sketch of it. In June 1842 I first allowed myself the satisfaction of writing a very brief abstract of my theory in pencil in 35 pages; and this was enlarged during the summer of 1844 into one of 230 pages, which I had fairly copied out and still possess.

The 'theory by which to work' was, of course, natural selection and Darwin would spend the rest of his life working on it in one form or another, although he would not make it public for another twenty years and then only when prompted to do so by the realisation that Alfred Russel Wallace had arrived at a similar theory from his own research.

This is not to say that Darwin did nothing with his theory during this period. In 1842, he felt confident enough in his research and his conclusions to write a short essay setting out his theory, which he would expand into a manuscript of 230 pages in 1844. This was important enough for him to make provision for it to be published by his wife in the event of his death and it would form part of the material read out at the Linnean Society in 1858, when natural selection was made public for the first time. In the meantime, Darwin embarked on a mammoth study of barnacles which, in the end, kept him occupied for eight years. After publishing four volumes on what some might think was an esoteric subject, he finally returned to natural selection. In 1856 he began writing his 'big book' on the subject, although the project was never completed because, two years later, he received the shock from

Wallace and began to write a shorter version which became *The Origin of Species*.

Evolution after the *Origin*

Darwin's Later Years

Ever since Darwin had moved out of London to Down House in the Kent countryside he had preferred the company of his immediate family, with the occasional visit from a close friend, to involvement in a wide and hectic social scene and, as he got older, this tendency became even more enhanced. He gave the impression of being at his happiest when in his study or when pottering around in his garden and greenhouse, particularly when engaged in one of the various botanical studies he developed to test out one of his ideas. The first book he wrote after *The Origin of Species* was published appeared in 1862 and must have come as something of a surprise to anybody expecting another sensational volume on evolution. Darwin wrote a treatise on one of his favourite families of plants – the orchids – and specifically on their fertilisation by insects. In characteristic style, he came up with a long and convoluted title, *On the Various Contrivances by which British and Foreign Orchids are Fertilised by Insects*. On this occasion, perhaps thinking that a book with Darwin's name on the spine would sell whatever the title, John Murray did not intervene.

In the years immediately after the publication of the *Origin*, Darwin began to grow orchids in his greenhouse, admitting he had become obsessed by them, and, although he would continue his letter-writing campaign in support of natural selection, he found that having something else on which to concentrate took his mind off the controversy surrounding evolution and relaxed him. In the second half of the nineteenth century botany and horticulture became extremely popular and fashionable in Britain and Darwin participated in this craze himself, corresponding with professional plant breeders and amateur growers, writing articles and letters to gardening magazines, and getting Hooker, the director of Kew Gardens, to send him interesting plants from around the world and introduce him to all sorts of contacts. At the time he was the most famous naturalist in the world and it is easy to imagine that those people he approached were only too happy to help in whatever manner they could. For Darwin, these various correspondents worldwide must have provided an antidote to the antipathy and vilification he had aroused with the *Origin*.

The orchid book was mostly about the adaptations shown by the plants in order to attract their insect pollinators and, in this respect, can be seen as a continuation of Darwin's use of examples in the *Origin* to illustrate how natural selection worked. The relationship between orchid and insect would now be described as an example of co-evolution, in which two species interact with one another, each exerting selective pressure on the other and resulting in the two becoming mutually dependent. One way of looking at Darwin's book, then, would be to think of it as a surreptitious attempt to further his evolutionary ideas in an innocuous little book ostensibly about nothing more than orchids.

The next book he wrote was about climbing plants and, needless to say, it concentrated on the adaptations shown by these plants. Then Darwin began work on a major book, *The Variation of Animals and Plants Under Domestication*, which, he said, took him three years of hard work to write, interrupted by numerous bouts of illness. It was published in two volumes in 1868 and in it he addressed one of the major problems highlighted by critics of his ideas on natural selection – the fact that he was unable to describe the way in which adaptations passed on from one generation to the next. Darwin called his hypothetical solution to this problem 'pangenesis' and it was one of the few occasions when he presented a theory which was not based on observation. It was also one of the few occasions when Darwin was completely wrong.

Darwin would return to this theory in his next book, *The Descent of Man*, published in 1871, although, as the title suggests, his primary intent was to extend the theory of evolution to include human beings. Twelve years previously, he had written in the *Origin* of his intention to do this, saying 'light would be thrown on the origin of man and his history', the only reference to humans in the entire book. By the time he actually got around to doing so, the subject was not nearly as controversial as it had been. Many scientists had come to accept evolution, at least to some degree. It was published by John Murray in two volumes and the content of the book can also be divided into two clear halves. In the first half, Darwin set out the case for humans to be placed in the tree of life, presenting an overwhelming quantity of evidence to show we are descended from what he described as 'some lower form'. He also tackled the controversial subject of race, expanding his view that all the races were descended from a common ancestor, a view taken for granted now, but contentious at the time. In the second half of the book, he introduced his theory of sexual selection, the evolutionary pressure exerted within a species during the process by which individuals selected a mate, as exemplified by the spectacular tail feathers of the peacock when compared with the drab colours of the female birds.

In the process of writing *The Descent of Man*, Darwin had intended to include a chapter on how emotions are expressed through facial expressions and body movements, a subject now regarded as having more to do with psychology than evolution. (At the time, psychology was only just beginning to be thought of as a discipline in its own right.) The quantity of material Darwin accumulated rapidly expanded until it became clear to him that it would be better published as a separate book. In the following year he published *The Expression of Emotions in Man and Animals*, a book that can be regarded as the last of his major works.

This is not to say that Darwin retired. Rather, he returned to his greenhouse and to the botanical studies in which he had previously been engaged and which he had been forced to neglect because of the demands of the intensive period of writing he had just endured. Although he was now 63, he still continued to publish papers and articles in various journals and magazines and was kept busy preparing new editions of his books, but he mostly confined his activities to his particular interests in plant breeding and amateur experimental botany.

Four more books on botanical subjects were published over the next eight years, covering such topics as insectivorous plants and the different forms flowers can take on the same species of plant. For what would become his final book he changed tack slightly, returning to a subject he had first considered more than 40 years previously. In 1837, in one of the first papers he had written which was not directly concerned with the *Beagle* voyage, Darwin examined the effects of the actions of earthworms on the soil and, in 1881, he published *The Formation of Vegetable Mould, through the Action of Worms*. The vegetable mould of the title was the dark coloured upper layer of soil in which there is a high proportion of organic matter, derived from decaying plant material, which is now generally called humus. When the book was first published, Darwin was worried whether anybody would buy a book on the subject of earthworms, although it actually sold well, becoming one of his best-selling titles.

In one of the huge number of essays Stephen Jay Gould wrote on and around the subject of evolution, he examined Darwin's last book, which has often been thought, perhaps with some justification, to be the eccentric project of an ageing man.[15] But Gould found the book to be much more than this. He saw it as a summation of a life's work, although he was not sure that Darwin consciously intended it to be read in this way. In the book Darwin explained how the upper layer of the soil is formed by the gradual but constant and ongoing sifting effect earthworms have on the soil as they swallow it, pass it through their intestinal tracts, where nutrients are absorbed, and eventually expel it as what are known as casts. The actions of a single earthworm are insignificant, but the overall result of the vast numbers of earthworms present in the soil over long periods of time is to keep churning the soil over, so that it is maintained in a relatively constant state.

The actions of worms can be seen as a metaphor for evolution as a whole, where almost imperceptible change is constantly occurring, with huge implications over long periods of time. If these changes can be observed and understood, then it becomes possible to project them back to understand the immense changes that have occurred over evolutionary and geological time, together with the processes which have driven these changes. Gould goes on to draw parallels between the book on earthworms and the first book Darwin wrote after he had published his account of the *Beagle* voyage, an account of how coral atolls and reefs are formed by the gradual accumulation of calcium carbonate, secreted by the coral polyps to form their exoskeletons. Here again is an account of a gradual natural process. All the books Darwin wrote between the book on coral and the book on earthworms, including the *Origin* and the *Descent*, can be seen in this light, as contributions to a lifetime's work intended to explain the workings of the natural world.

Towards the end of 1881, Darwin's already fragile health began to deteriorate further. He became increasingly weak and began to suffer from pains in his chest. Emma Darwin, who had patiently nursed him through many periods of ill health during the entire span of their married life, effectively became his full time carer, spending many hours reading aloud to him or playing backgammon, particularly after he grew too weak to take walks in the garden. As well as his wife, he was kept company by his sons and daughters and numerous grandchildren and, typically of him, was treated by at least four different doctors. In the afternoon of 19 April 1882, after suffering what was, in all likelihood, a heart attack during the previous night, he died at the age of 73.

A week later he was buried in the nave of Westminster Abbey, near Isaac Newton's grave, although Emma Darwin had hoped he would have been buried closer to Charles Lyell, who had died in 1875 and was also interred in the abbey. Darwin had expressed his wish to be buried in the churchyard near his home in Downe, but his friends, including Hooker and Huxley, campaigned for him to be interred in the abbey because of his great contribution to science. It was an irony Darwin himself, who had shown a keen sense of humour throughout his life, may well have appreciated, as he had done as much as anybody to undermine the position of the Church in British society.

Darwin and Society

Almost as soon as the *Origin* was published, Darwin's theories of natural selection and the struggle for existence began to be extended into other areas. The application of a theory which saw competition between individuals for resources in the natural world as a key factor in the continuing development of the system to the commercial capitalism of Victorian Britain was obvious. After all, Darwin had been influenced by the writings of the economist Adam Smith in the first place, but the extension of Darwin's work certainly did not stop there. One of the most influential thinkers of the period was the philosopher and social theorist Herbert Spencer, although his reputation would decline dramatically in the later years of the nineteenth century and particularly after his death in 1903. Put simply, Spencer saw human society as a natural phenomenon which went through a series of stages as it developed and, as such, was subject to the same evolutionary processes which shaped the natural world.

In the model of human society he developed, Spencer proposed that the mechanism of change was Lamarckian, with societies developing by passing on what they had learned to future generations, rather than through Darwinian natural selection. However, he incorporated into his model the idea of the struggle for existence, in which he envisaged the 'fittest' individuals in a society naturally rising to the top, and he coined the phrase 'the survival of the fittest' to describe this process. Put in these terms, the evolution of society could be used as a justification for the imperialist policies of Victorian Britain, with the strong colonising the weak, and for the exploitation of the poor by the wealthy in the expanding industrial economy of Britain. This allowed the appalling working conditions prevalent in some factories and the terrible living conditions in urban slums to be explained away as the inevitable result of evolution. Needless to say, the supporters of such ideas came from the upper echelons of society, who saw themselves as being at the top of the evolutionary ladder.

Spencer was probably the most influential writer on what would later be termed social Darwinism, but he was far from being the only one. The ideas encompassed by the term became the subject of a huge amount of debate, which would continue well into the twentieth century. To give just one example, in 1899 Rudyard Kipling published the poem *The White Man's Burden* in which he presents non-western peoples as childlike and undeveloped, and implies that the western powers are under an obligation to colonise them for their own benefit. Although the poem can be read as a satire on imperialistic attitudes, the title became synonymous with the perceived right of powerful Western countries to dominate the rest of the world in order to facilitate their evolution into more developed societies.

It is difficult to gauge how much sympathy Darwin himself had with the views expressed by Spencer and other writers on the evolution of society. He was certainly well aware of this strand of thought, and wrote in his autobiography of his admiration for Spencer. Yet he would also say that Spencer's work had not been of any use to him. As mentioned in the previous section, Darwin tackled the subject of race in *The Descent of Man*, in which he expressed his belief, explicitly racist but widely held at the time, in the superiority of the white race. In the section on sexual selection, he would also articulate another common view – that men were superior to women – and this went on to make him deeply unpopular with the emerging women's suffrage movements of the late nineteenth century. But Darwin also thought that people possessed an innate moral conscience and a sense of sympathy for others which strengthened society, lessening the tendency towards conflict implicit in a view of society which emphasised the role of competition above all else. With the exception of remarks on these contentious subjects, which Darwin thought essential in a book about human evolution, he confined himself in the rest of his writings to what he considered to be the scientific subjects of botany, zoology and geology.

Darwin may not have been keen to venture far into the subject of the evolution of society, but his cousin Francis Galton was not so reticent. He was one of the last of the Victorian polymaths, an independently wealthy gentleman who, not unlike Darwin, used his independence and connections to engage with whichever subjects took his fancy. The difference between the two men was that Darwin mostly confined himself to the study of the natural world, while Galton moved between travel and exploration, statistics, meteorology, psychology and anthropology, making important contributions to the advancement of a number of these disciplines. However, today he is remembered more for his writing on social issues, particularly for developing what would become the notorious field of eugenics, a strand of social theory concerned with the study of inheritance of character traits in people, with the objective of developing strategies for improving the fitness of human populations.

After reading the *Origin* shortly after it was published, Galton applied Darwin's mechanism of evolution – natural selection – to society, thus achieving a kind of circularity since Darwin had been influenced by Malthus's work on population to develop his theories in the first place. Galton reached the conclusion that any measures taken to alleviate the deprivations then common in Victorian society would prevent natural selection from taking its course, leading to a decline in the overall fitness of the general population. Any actions taken by the Government or by charities and religious groups which aimed to reduce poverty, homelessness or ill health among the lower classes, or sought to improve living standards, working conditions and levels of education would, according to this thesis, only serve to allow people who were subjected to these deprivations to continue to have more and more children, swamping Britain with the unfit and depraved.

Galton considered almost every trait to be inheritable and, in the first book he wrote on the subject after reading the *Origin*, he examined the concept of the innate genius, using the statistical analysis he had developed to show that the children of the highly intelligent were more likely to share that trait than those with less intelligent parents. It was a short step from this conclusion to arguing that the poor and destitute had fallen on hard times because of an inherited lack of intelligence, along with a predisposition towards laziness and a lack of morals. He also claimed that the lower echelons of society, particularly those exhibiting various forms of depraved behaviour, tended to have the largest families. So, as time went on and these traits were passed from one generation to the next, society would gradually degenerate.

The solution to this problem Galton favoured was what would later become known as 'positive eugenics', which entailed encouraging the intelligent, honest and hard-working people in the population to have larger families to counteract the greater fecundity of the undesirable elements in society. An alternative solution was what would become known as 'negative eugenics', which proposed subjecting those people thought to be unfit to interventions in their lives which would reduce the number of children they were having. This was suggesting, in essence, the adoption of a system of human breeding similar to that used by livestock breeders to select their best animals to carry on the line and improve the pedigree. It was not a method advocated by Galton, who was, like Darwin, a supporter of the Whigs, a political party whose policies were based on the premise that the Government interfered in peoples' lives as little as possible. But, as concern over the state of the country grew, so too did support among other social commentators for the adoption of policies based on negative eugenics.

These ideas about engineering a better society may now seem like the ramblings of a crackpot, but they were certainly not seen that way in the later part of the nineteenth century and the early part of the twentieth. Galton was a highly respected scientist and his theories on society were given further respectability and rigour through their association with Darwinian evolution. Darwin was not overly impressed by Galton's thoughts on the inheritance of genius, saying that he thought achievements came as much through zeal and hard work as through intelligence or any innate tendency towards genius. He died before Galton set out the complete principles of eugenics in a book published in 1883 and, although Darwin was not above repeating racial and sexual stereotypes himself, it is hard to see him endorsing his cousin's ideas. In *The Descent of Man*, he had expressed a certain amount of agreement with Galton's earlier theories, but he had qualified them by writing that he thought social instincts and sympathy for others arose through natural selection, which would lead to the development of a social conscience and the strengthening of society.

The eugenics movement begun by Galton began to take off around the turn of the century, encouraged by population studies showing that the numbers of the underprivileged in society were increasing and by reports of a general increase in unfitness. This growing unfitness, taken as a sign of a deteriorating society, was clearly illustrated when a large number of potential recruits to the military during the Boer War (1899–1902) had to be rejected on medical and physical grounds. The Eugenics Society was formed in 1907, with Leonard Darwin, one of Darwin's sons, acting as its president for a number of years, and the National Eugenics Laboratory was established at University College London to further the study of the degeneration of society. An outcome of this research was the adoption of a negative eugenic strategy by the British Government in 1913 when the Mental Deficiency Act was passed, which separated the 'mentally defective' into categories and made provision for their incarceration in hospitals and asylums so that they would not be able to have children. The loss of a generation of what were thought of as the fittest young men in society in the trenches of the First World War served to increase the mood of despondency further.

By this time eugenics had spread to continental Europe and America where, if anything, it was taken up with even more enthusiasm than in Britain. Laws allowing for the compulsory sterilisation of mental patients and criminals were passed in more than 30 American states, while, in Germany, eugenics and Darwinism in general were used to justify the supposed superiority of the Germanic races. When the Nazi regime gained power in Germany in 1933 numerous laws were passed which were based on eugenic principles and were aimed at improving the racial and physical characteristics of the population. These included widespread sterilisation programmes and the prevention of marriages between Germans and Jews. They were only the forerunners of the obscene policies which would follow, culminating in the 'final solution', the euphemism adopted by the Nazis for the genocide of six million Jews, and the wholesale murder of any other group considered undesirable, including Slavs, the Roma, homosexuals and the mentally ill.

In the aftermath of the Second World War, and in the full knowledge of the Holocaust, the ideas of improving society presented by eugenics largely fell from favour, although some eugenic practices would persist for a considerable time. The last compulsory sterilisation was carried out in America as late as 1981. The extent to which the Nazi regime actually based their genocidal policies on the ideas of eugenics is hard to gauge. It may just have used them as an excuse and it is probable that, even without eugenic theory, it would have gone ahead with those policies anyway and found another justification.

The misappropriation of scientific theories to justify all sorts of odious practices has occurred on numerous occasions. Stephen Jay Gould wrote of natural selection:

> This theory has a history of misuse almost as long as its proper pedigree. Claptrap and bogus Darwinian formulations have been used to justify every form of social exploitation – rich over poor, technologically complex over traditional, imperialist over aborigine, conqueror over defeated in war.[16]

This quote is typical of Gould, who was not a man to mince his words over what he considered to be the perversions of Darwinian principles. He was also highly critical of more recent research in the related fields of sociobiology and evolutionary psychology, seeing them both as examples of biological determinism, the idea that the form and character of an individual organism is entirely the result of the expression of its genes rather than the influence of social or environmental factors. In this sense, the continuing debate over sociobiology and evolutionary psychology can be seen as a continuation of the 'nature versus nurture' argument, a phrase first used by Francis Galton in the 1870s.

Sociobiology concerns itself with the application of evolutionary theory to the explanation of animals' social behaviour. The word was first coined in the 1940s and came into widespread use after the publication in 1975 of *Sociobiology: The New Synthesis* by Edward O. Wilson, an eminent evolutionary biologist who has specialised in the social behaviour of ants. The use of evolutionary theory to describe the extraordinary and complex behaviour of ants has not been particularly controversial, but attempts to explain human behaviour using a similar set of principles ignited a furious row. At the extremes of the argument are the opposing positions of complete biological determinism on the one hand and, on the other, the idea of the *tabula rasa*, a Latin phrase meaning 'blank slate' which has been used to illustrate the argument that human beings are born with no innate mental resources, and that their knowledge is entirely built up through the accumulation of experiences of the outside world. No serious evolutionary biologist or psychologist would adopt a position at either extreme of the argument, and almost all commentators recognise that there are complex interactions between genes, culture and environment. The debate has revolved around the degree to which different factions have leaned towards one side of the argument or the other, with each side accusing the other of adopting positions which reflect the political and cultural attitudes of the individuals concerned rather than the scientific evidence.

Evolutionary psychology attempts to explain mental functions, such as an emotional response to a particular situation or the innate ability of young children to learn a language, as adaptations to the environment in which human beings evolved many thousands of years ago. Critics of this approach follow a similar line to critics of sociobiology and, in particular, accuse evolutionary psychologists of being overly creative in developing evolutionary explanations for psychological traits. These explanations are sometimes called 'just so stories' by opponents of evolutionary psychology, who consider them to be unprovable and thus lacking in scientific rigour.[17]

Similar accusations have been levelled at the concept of 'memes', a term first used by Richard Dawkins in 1976 in *The Selfish Gene*. A meme, according to Dawkins, is a unit of cultural inheritance which can spread through society and change over time, analogous to the gene in biological inheritance. He gives the examples of ideas, fashions, songs and, as he writes, 'ways of making pots or of building arches'. In typically controversial fashion, Dawkins gives religion as an example of a meme, although he might just have well chosen Darwinian evolution or, indeed, anything generated in the human brain that has spread through culture.

Dawkins would later say he introduced the concept initially only as another example of a replicator, although he would describe memes as unreliable in comparison with genes because they can be changed as they are passed on. For example, subtle differences appear as a fashion for a particular type of clothing spreads and different people adopt the fashion but modify it to suit themselves. Needless to say, the concept of memes has proved controversial. Perhaps the most serious criticism is that memes, unlike genes, are metaphorical constructs, and so it is impossible to test for their existence. The whole concept of the meme, according to critics, is not a scientific one.

Whether it is a part of science or not, memetics, as the study of memes is called, does illustrate the extent to which any application of Darwinian evolution to human society and culture is likely to be contentious, in the same way that the application of evolutionary principles to humanity in general was in Darwin's day. Darwin himself used the anatomical similarities between human beings and the great apes to show how closely related we are. Subsequent discoveries of the fossilised remains of early hominids, such as *Homo erectus*, *Homo habilis* and the Neanderthals (*Homo neanderthalensis*), provided more supporting evidence and also corroborated Darwin's theory that humans evolved in Africa. Today DNA evidence shows we are actually much more closely related to the apes than had previously been thought, and that humans and chimpanzees diverged only about five million years ago.

The past misuse of Darwinian evolution to justify various forms of exploitation and subjugation, including sexism, racism and the atrocities of the Nazi regime, means that extreme caution is required in any attempt to extend evolution beyond its use in describing biological systems, and that an appreciation of the consequences of previous applications of social Darwinism is important. With such precautions in mind, it is possible for Darwinian evolution to provide powerful tools for the furthering of our understanding of ourselves.

Evolution Today

By the time of Darwin's death in 1882, evolution had become an established theory, widely accepted within the scientific community even if still controversial in some sections of society. But the version of evolution put forward by Darwin, of variation by natural selection, was not always the favoured model. Rival theories were developed and older theories, particularly Lamarckism, found acceptance in some quarters. Advances in evolutionary theory were hampered in the last decades of the nineteenth century by an inability to make any headway in uncovering both the mechanism of inheritance and the process by which variation occurred.

In these two decades, biology began to become a much more laboratory-based science, as it remains today, using techniques developed in physics and chemistry to investigate the cellular and molecular basis of life. In the 1880s, the cell nucleus was found to be where hereditary material was held and the role of the chromosome as the carrier of information between parents and offspring was demonstrated not long afterwards. In 1900, the experiments Gregor Mendel had carried out on peas in the 1850s and 1860s were rediscovered and their importance in terms of inheritance established.[18] This would be the starting point for the development of genetics which was, initially at least, considered to be an alternative to Darwinian evolution.

The divisions between genetics and evolutionary biology and between the work carried out in the laboratory and in the field continued throughout the early years of the twentieth century, with a limited amount of communication between scientists working in different areas. It would be the 1930s and 1940s before the different disciplines began to come together, resulting in what is known as the 'modern synthesis' of genetics and evolutionary biology. The publication in 1942 of the book *Evolution: The Modern Synthesis* by Julian Huxley, the grandson of Thomas Huxley and one of the many people involved in the work, has been regarded as the moment when the central tenet of modern evolutionary biology was first expressed. The two main sticking points of Darwinian evolution were resolved by genetics. Variation in the phenotype, the outward appearance of an organism, was shown to be possible through changes in the genotype, the genetic compliment of an organism, by the process of mutation, and the mechanism of genetic inheritance showed how character traits and variations could be passed on from one generation to the next. A huge body of research has been undertaken to uncover further details and to refine and correct the model, but the modern synthesis still remains the fundamental building block on which modern evolutionary biology rests.

In 1953, James Watson and Francis Crick described the structure of DNA (deoxyribonucleic acid), the molecule known to carry genetic information, as the now famous 'double helix'. Because of this discovery it became possible to explain how genes hold the information of inheritance, which is coded in sequences of so-called base pairs, how they replicate and how they pass information from one generation to the next. Between 2000 and 2003 a number of different research groups published papers describing the human genome, the complete sequence of base pairs which make up human DNA, leading to estimates that there are about 25,000 genes in the genome. The implications of this knowledge, for medical science as well as for evolutionary biology, are enormous, although the research has also uncovered the extraordinary complexity of gene expression, the method by which the information held by a gene is translated into a trait. Rather than a single gene – a particular stretch of DNA – being translated and thus being expressed directly in the phenotype, in the majority of cases a network of genes is involved in the expression of a particular trait, and various regulatory factors determine how that expression occurs.

Huge advances have been made in the field of evolutionary biology as well in genetics since the modern synthesis. The range of research has been far too wide for even a representative sample to be incorporated in this book, so one example is here used as an illustration.[19] In a paper published in 1972, Niles Eldredge and Stephen Jay Gould addressed what was one of the fundamental problems of evolution. The fossil record, although it had been hugely extended since Darwin's day, still did not show the gradual changes that would be expected from the progress of evolution. Instead the record showed long periods in which nothing much appeared to happen, then sudden jumps when large numbers of new species began to appear. In 1972, this was explained by using much the same argument Darwin had used in *The Origin of Species*. It was the result of the chance nature of the creation of fossils, and the vast majority of evolved species simply did not appear in the record before becoming extinct.

Eldredge and Gould challenged this established idea by proposing that, even though the fossil record was not complete, it still presented a good overall picture of evolution. They used a model for evolutionary change they called the 'punctuated equilibrium' in which species remained much the same throughout their history, but had their stasis occasionally interrupted by short and rapid periods of change. Another way of putting this would be that the established equilibrium was occasionally punctuated by rapid moments of evolution, leading to the abrupt changes in the species found in the fossil record. These punctuations, they proposed, would occur when a small number of individuals of a species became isolated from the general population and, under a different set of environmental conditions, rapidly evolved into new species, as had happened on the Galapagos Islands with Darwin's finches.

Critics of punctuated equilibrium have seen it as the extension of Marxist ideology, in which a society can only be changed by revolution, into the field of evolutionary biology. Those who favour the theory of evolution occurring gradually point to the continuing process of genetic mutation which is random but, over a long period of time, statistically predictable as the main agent of variation. However, these two different views of how evolution works do not necessarily have to be mutually exclusive. There is no reason to suppose that evolution can only follow one pathway, the natural world having shown itself to be incredibly diverse and multi-layered.

The Controversy Continues

Evolution and Christianity

The opponents of evolution in Darwin's day, at least those whom he took seriously, were some of the most established scientists of the time who held to the predominant view that the natural world was the creation of God for the benefit of humanity. In the years after the publication of *The Origin of Species* in 1859, many scientists began to accept some form of evolution, although not necessarily the version put forward by Darwin. Darwin's work can be seen as contributing to an ongoing process in which the sciences became separate from religion, one that can be dated back to Copernicus in the early part of the sixteenth century, who was the first person to show that the Earth was not the centre of the Universe but a small planet revolving around the Sun.

The Christian Church took some time to accept the Copernican revolution, as Galileo would find out to his cost when he was convicted of heresy for supporting the heliocentric theory and spent the rest of his life under house arrest. It would be another hundred years before Copernicanism was accepted by the church, after Isaac Newton made it impossible for anyone who seriously engaged with the subject to adopt any other view, and it rapidly became the established scientific viewpoint, only to be questioned by cranks. Darwinism, it is fair to say, has not received such a rough ride from mainstream Christianity, and most denominations have eventually reached an accommodation of one sort or another with evolution which has allowed believers to follow their faith while accepting the advances of science. The Roman Catholic Church, for example, adopted a position on evolution in 1950 with the issue of a papal encyclical by Pope Pius XII, which accepted that the hypothesis of the evolution of the human body could be reconciled with Catholic doctrine, while also stating that the soul was created by God. In 1996, Pope John Paul II endorsed the encyclical of his predecessor and went on to recognise that a significant quantity of research since the 1950s had advanced the theory to the point where it could no longer be doubted.

The research referred to by Pope John Paul II can be seen as a direct result of the work done in the 1930s and 1940s to bring about the modern synthesis, which placed evolution at the core of the biological sciences. Evolution has actually been observed in both the laboratory and in the field, including in what has become a classic study of Darwin's finches by Peter and Rosemary Grant. Over a period of thirty years of research, the Grants have documented evolutionary adaptations in the finches as a response to changing environmental conditions.[20] Other examples include the discovery of intermediaries between different classes of animals in the fossil record (such as the tiktaalik fish found in northern Canada in 2004 which shows some characteristics of an amphibian), the development of bacterial resistance to antibiotics and the extremely close similarities between the DNA of human beings and the great apes.

The list and range of examples goes on and on and, taken together with the fact that no single piece of evidence has ever been found to refute evolution, you might think they would be enough to convince anyone that it is a genuine and observable phenomenon. The evidence has been good enough for the Pope but, if anything, objections from some other Christian denominations appear to be growing. (Of course, it may just be that the objectors are making more noise on the subject than those who accept evolution.) Paradoxically the majority of these religious groups objecting to evolution come from America, the country in which many of the recent advances in evolutionary biology have been made and where secularism is enshrined in the constitution. This has been a consequence of the rise of Christian fundamentalism, a movement which began in the early twentieth century and has remained almost exclusively American. It consists of an alliance of Christian groups who share a set of principles and beliefs, who believe that the Bible is literally true and object to what they perceive as an increasingly liberal and permissive society.

Darwinian evolution was taken by Christian fundamentalists as a direct challenge to their faith, as it contradicts the description of the Creation in Genesis, and so it has become one of the main targets of their anger. As we shall see later, this has often found expression over the years in attempts to get the teaching of evolution banned in schools. As the fundamentalist movement grew it began to find a political voice, perhaps following the example of other religious groups such as the Jewish lobby, in an attempt to gain influence and further its evangelical purposes.

While this was happening, science in general, including evolutionary biology, became increasingly specialised, with findings published solely in academic journals and papers written in such a way as to be almost incomprehensible to anyone who had not themselves specialised in that particular discipline. With the exception of the contributions of a small number of scientists, such as Richard Dawkins and Stephen Jay Gould, who have considered engagement with the general public an essential part of what they do, much of modern science occurs behind closed doors and it keeps itself separate from the rest of society. Not without some justification, non-scientists have become not only baffled by what science is doing, but also suspicious of it and wary of its conclusions.

In their objections to Darwinism, Christian fundamentalists have forced evolutionary biologists to engage with their arguments whether they want to or not, which has often resulted in a futile debate between two sides unlikely even to listen to the other side's point of view never mind accept its argument. People with strongly held religious views are hardly likely to change their minds when presented with the evidence of evolution, however convincing it may be, while evolutionary biologists are equally unlikely to undergo a religious conversion after being exposed to fundamentalist beliefs.

In an attempt to resolve the never-ending and, on both sides, unwinnable debate between science and religion, Stephen Jay Gould came up with the idea of non-overlapping magisteria, or NOMA, which he developed after reading Pope Pius XII's encyclical. Essentially Gould suggested that science and religion occupied two different spheres of human thought. Science covered the empirical realm of knowledge that can be obtained through observation and experimentation, while religion, Gould wrote, 'extends over questions of meaning and moral value'. This is much the same as the Pope saying that the human body is open to investigation, but the soul belongs to God. Rather than being a resolution to the problem, it gave the impression more that Gould had grown tired of constantly having to defend evolution and that he wanted to disengage himself from any further debate with Christian fundamentalists. Scientists, after all, should be under an obligation to consider the moral and ethical dimensions of their work, while the deeply religious would surely argue that their faith covers all aspects of the human condition.

Richard Dawkins has apparently adopted the opposite approach to that of Stephen Jay Gould, one which uses the method of full frontal attack rather than appeasement. This has led to him being described by some people as Darwin's Rottweiler and as a militant, or even a fundamentalist, atheist. However, in some ways, his aims could be interpreted as similar to those of Gould. In setting out his views in straightforward and unequivocal language, as

Dawkins has done in *The God Delusion*, published in 2006, he may have been attempting to get beyond the persistent questioning of them so that he could return to his primary concern of evolutionary biology. If this was the case, then it has not worked. *The God Delusion* has become an international bestseller and Dawkins probably the most famous atheist in the world, much more in demand for his views on religion, which are endlessly discussed, than he ever was for his ideas on evolution.

In *The God Delusion*, Dawkins sets out his position as an atheist and secular humanist, a philosophy which promotes reason, ethics and justice while rejecting anything to do with the supernatural and spiritual, and, in the process, he performs a demolition job on all forms of religious belief. *The God Delusion* was by no means the only book published on atheism and the deleterious effect of religion at this time[21] and, accompanied by numerous books rebutting Dawkins's anti-religious views, these anti-religion volumes have become something of a publishing phenomenon. It would seem that even in what are sometimes described as the post-Christian societies of Western Europe as well as in the supposedly secular country of America, religion remains a fascinating subject for many people, even if that fascination sometimes concentrates on the insidious influence religion is said to have.

Monkey Trials

In *The God Delusion* Richard Dawkins expressed some forthright opinions on all manner of religious practices but he was particularly critical of what he described as the indoctrination of children with religion in faith schools, disagreeing with it in the strongest possible terms and even going so far as to compare it with child abuse. Most children, it is probably true to say, first experience religion through their parents rather than through school but the continuation of a child's religious education in school must, nevertheless, be an important formative experience and deeply influence the ways in which they view religion in later life. Christian fundamentalists in America have certainly recognised the importance of education, but, unlike in Britain, where Religious Education remains a compulsory subject in state schools, the First Amendment of the US Constitution prohibits any religious teaching in state schools throughout the country.

In multicultural Britain, religious education often takes the form of a non-critical examination of the religions of the world and, although it would be theoretically possible for creationism to be introduced into this study, such a move would almost certainly prove highly controversial. Such a scenario does not apply in America, since religious education is not allowed in the first place, but some Christian fundamentalist groups have attempted to get around the prohibition by setting up their own schools. These schools are usually funded by donations rather than by state taxes and, at least in most states, can include religious classes in the curriculum. However, the primary motivations behind the formation of these schools appear to be as much to do with objections to the teaching of evolution in science classes as they are to do with providing separate religious education, which often existed already through such institutions as the Sunday School.

An alternative tactic adopted by fundamentalists has been to attempt to establish laws within particular states which either ban the teaching of evolution in state schools or that make provision for some form of creationism to be taught alongside evolution in science classes to provide an alternative view to what they see as an unproven theory. This has invariably led to the issues involved being tested in the law courts, and goes back as far as what has become known as the Scopes Monkey Trial of 1925, when John Scopes, a teacher in the small town of Dayton, Tennessee, was prosecuted after breaking a recently enacted state law, the Butler Act, which forbade the teaching of evolution in Tennessee state schools.[22] The American Civil Liberties Union had offered to finance the defence of any teacher prepared to break this law. Scopes, with the backing of some local citizens who thought the trial would bring publicity to the town, accepted the challenge, although he was never actually called as a witness for his own defence at the trial, probably because, as he would later admit, he could not remember for certain if he had actually taught evolution in the few science classes he had taken in Dayton. After some dubious tactics by the defence lawyer, which appeared to have more to do with showboating than with a serious attempt to win the case, Scopes was found guilty and fined $100. The defence appealed the case to the Supreme Court, where they hoped it would be found unconstitutional, a ruling that would lead to the removal of the Butler Act from the statute books. However, the case was thrown out on a technicality and the act remained as law in Tennessee until it was repealed in 1967.

In 1968, in another case, this time in Arkansas, the Supreme Court ruled that banning the teaching of evolution really was unconstitutional, and this forced the creationists to change their strategy and attempt to get 'creation science', as they called it, taught alongside evolution. A case brought in Louisiana in 1987 as a result of the passing of a state law requiring both evolution and creation science to be taught in science classes led to a court ruling that anything to do with creationism, whether or not it was presented as a science, amounted to a religious doctrine. Therefore it could not be taught in state schools as this would be unconstitutional. This ruling led directly to the establishment of the intelligent design movement, as creationists essentially rephrased their argument to exclude any direct reference to God or the Creation, and replaced their main line of reasoning, the argument from design, with more scientific sounding phrases. They now spoke of such concepts as 'irreducible complexity', the idea that a biological system composed of a number of different parts, such as the immune system, could not have arisen as a result of evolution because all the parts would have had to have evolved at the same time for it to have worked and to have resulted in an inheritable advantage. The only way such systems could have arisen, according to the theory at least, was if there was an intelligent designer who, in effect, put the parts together.

The only real distinction between irreducible complexity and the argument from design is that the intelligent designer is not actually identified as God. A court case brought in 2005, known as the Dover Trial, acted as a test case for intelligent design and, unfortunately for the creationists, it was found to be creationism by another name. The case was brought by the parents of children attending a school in Dover, Pennsylvania, after the school board voted to include intelligent design in science classes. The loss of the case constituted a major setback for the intelligent design movement. In the published decision, the judge wrote:

> To preserve the separation of church and state mandated by the Establishment Clause of the First Amendment to the United States Constitution, and Art. I.3 of the Pennsylvania Constitution, we will enter an order permanently enjoining Defendants from maintaining the ID Policy in any school within the Dover Area School District, from requiring teachers to denigrate or disparage the scientific theory of evolution, and from requiring teachers to refer to a religious, alternative theory known as ID.[23]

In the face of such a stinging rebuke from the Supreme Court, many people, it might be thought, would have given up their attempts to, in effect, subvert the US Constitution but the people involved in the intelligent design movement are nothing if not persistent. Realising that there is no future in continuing to campaign for intelligent design to be taught alongside evolution on scientific grounds, their latest strategy is to avoid the mention of intelligent design but to promote the right of teachers to introduce alternative explanations into a discussion of evolution as an academic freedom. At the time of writing, in July 2008, a law has been voted onto the Louisiana statute books by the state legislature, called the Science Education Act, which allows teachers to present non-scientific theories alongside science as a means of promoting critical thinking skills. This is a much more subtle approach than any previous attempts to introduce creationism into state schools and one that does not in any way mention a religious motivation for doing so. The crux of the matter is a clause in the law which allows for the use of supplementary material in classrooms to help with discussions on scientific subjects. This offers a roundabout way, opponents of the law argue, of getting textbooks putting the creationist argument into schools in such a way as not to infringe the US Constitution. Although it has not happened yet, it looks as if another lengthy legal battle will ensue to test this law and continue the argument between scientists and creationists in America for a number of years to come.

In the event of this latest strategy actually succeeding and creationist literature being allowed into American schools, it has the potential to backfire on its advocates. Exposing an argument which attempts to use science (or, at least, scientific terms) to explain a theory which relies on faith rather than reason for its credibility, to critical thinking in the classroom, or anywhere else for that matter, is asking for trouble. An unbiased comparison between creationism and evolution in terms of scientific veracity hardly seems possible in America at the moment but, if it were to happen, the more likely outcome would surely be the undermining of the faith necessary to sustain a belief in creationism rather than damage to evolution. Perhaps the time has come to put the issues to the school children of America, allow them to apply their powers of critical thinking and let them make up their own minds.

A Grandeur in this View

It is impossible to know now what Charles Darwin would have made of the controversy over evolution that continues in America today but, with a little imagination, it is possible to speculate that he would most likely be astonished at the nature of the argument. It was one with which he was very familiar himself. It had been expressed in 1802 in the book *Natural Theology*, in which William Paley put forward the watchmaker analogy for the teleological argument, which is pretty much the same argument as that from design and irreducible complexity. Twenty-five years previously David Hume had taken this argument to pieces in his book *Dialogues Concerning Natural Religion*, although, in the absence of any satisfactory alternative explanation, he was forced to conclude that there was no other choice than to accept the argument from design.

Towards the end of the eighteenth century a number of naturalists and philosophers developed evolutionary theories, most of which included the idea of the inheritance of acquired characteristics in one way or another, most convincingly articulated by Jean-Baptiste Lamarck. None of these theories stood up to a rigorous examination and none was widely accepted. In Britain evolutionary ideas were also thought to be dangerously radical and associated with the disorder and godlessness of revolutionary France. Anybody expressing such views was considered a threat to the established order of society, in terms of the strictly defined hierarchies of both the Church and the ruling classes.

In describing the mechanism of natural selection, Darwin made the argument from design redundant and, once the initial controversy had died down, it became apparent that the advance in thinking he had made did not threaten society. Some people indeed used it to justify current social inequalities in terms of the 'survival of the fittest'. Put aside the social and political consequences of evolution, which he would surely have seen as the misuse of a scientific theory, and it is reasonable to assume that David Hume would have accepted Darwin's theory as a rebuttal of the argument from design. It is even possible to imagine William Paley, an intelligent and erudite man to judge from his various writings, engaging with the theory, perhaps in the same way that Darwin's friends Charles Lyell and Asa Gray, both deeply religious men, accepted natural selection and yet found a way of reconciling it with their beliefs.

The extent to which Darwin himself had to accommodate his evolutionary principles within his own beliefs is impossible to know because, throughout his life, he remained very unwilling to discuss his faith or, perhaps more likely, the lack of it. He may have felt that by being open about his thoughts on religion, he would only add to the controversy surrounding evolution. Unlike Thomas Huxley, he was not a man who enjoyed the cut and thrust of vigorous debate, generally doing whatever he could to avoid direct confrontations. The thought of upsetting his wife, who maintained her own religious beliefs throughout her life, may also have encouraged Darwin to remain silent about his views on religion. He may also have considered his beliefs his own business rather than the concern of anybody else.

A few years before he died Darwin wrote a short piece, not intended for publication, reflecting on his religious beliefs in which he said, 'I gradually came to disbelieve in Christianity as a divine revelation', but he does not go on to elaborate very much on how this happened. A great deal of academic endeavour has gone in to attempting to pinpoint the exact moment he lost his faith. The consensus view is that any remaining religious feeling he may have had disappeared when his ten-year-old daughter Annie died in 1851. In truth, if a person does not wish to share their thoughts with the outside world, it is all but impossible to reconstruct them from their actions, however convincingly these actions can be put together to form a coherent story. Although Darwin wrote in his reflections on religion about being a believer in his younger years, the overall impression he gives in his writing is that he was not particularly concerned with religion at any stage in his life. Both his father and his grandfather are usually described as freethinkers, something of a euphemism for those who did not believe in

God at a time when it could be socially disastrous to admit openly to atheism, and it would take no great stretch of the imagination to see Darwin as following in this family tradition.

Whatever the extent of Darwin's belief or non-belief, he was not about to allow it to interfere with his work as a scientist. He dedicated his life to the 'one long argument' he wrote about in *The Origin of Species*, in which he presented the case for evolution in a way that had to be taken into account by anybody coming after him, whether scientist, philosopher or religious thinker, who studied the natural world or humanity. It remains possible to oppose Darwinian evolution, but it has not been possible to ignore what Darwin said.

By articulating his theories of evolution Darwin did more than anybody before or since to further our understanding of the natural world. In showing the place of human beings within this world – rather than over and above it – he has also allowed us to better understand ourselves. Some see the Darwinian view of humanity as a bleak one, in which there is no meaning or purpose to life, but an alternative viewpoint would be that Darwin attempted to describe the world as it really is rather than how we would like it to be. He was far too polite to express himself in such a way, but the essence of what he was saying is that you can believe whatever you want to believe, but this is what it is really like.

In the five years Darwin spent circumnavigating the globe on board the *Beagle*, he visited some of the most beautiful and extraordinary places on the planet, seeing for himself the complexity and diversity of life in the Amazonian rainforests, the unique and unusual species inhabiting the volcanic islands of the Galapagos, and the desolate and windswept landscapes of Tierra del Fuego. He rode across the vast expanses of the Argentinian pampas, climbed in the Andes and collected specimens of the teeming life to be found in coral reefs and kelp forests. Back in England, he spent eight years studying barnacles and many more growing orchids and investigating the action of earthworms in his garden, all the time working on his theory which explained how all of it, from the earthworms to the rainforests, had come to be as it was. As he wrote in the final paragraph of the *Origin of Species*, 'there is a grandeur in this view of life' and there was also a grandeur in the life of the man who gave us the chance to see it.

Notes

1. The Darwin Correspondence Project has published 16 volumes of letters so far, collecting all the known letters up to 1868. Many of these letters are also available online at darwinproject.ac.uk Much of Darwin's other writing has also been published, details of which are in the bibliography, and can also be found at darwin-online.org.uk
2. This extract is taken from Buckhardt and Smith (1991). The theories of Darwin and Wallace were not actually as similar as Darwin claimed them to be, the main difference being that Wallace described competition between species as the main factor leading to change, while Darwin emphasised competition between individuals of the same species. In later life, Wallace could not accept that the human mind could have arisen by evolution alone and became interested in spiritualism, both factors which may have contributed to him becoming the 'forgotten man' of evolution, as he is often described.
3. Unfortunately for Thomas Bell, he is now only remembered for this comment, which is quoted in most biographies of Darwin and is here taken from Browne (2002).
4. The name of the village was changed from Down to Downe after Darwin moved into Down House. He never bothered to change the name of his house.
5. At the time of writing in July 2008, Peter Harrington Books in London have a first edition of *The Origin of Species*

in stock, which they describe in their catalogue as being the most important book ever published. It is one of the original 1,250 printed by John Murray in the first issue and is for sale at £95,000.

6. While travelling in South America, Darwin would later write, he always took a small edition of Milton with him.
7. As well as not using the word 'evolution' very much in the first edition, Darwin also did not use the phrase 'the survival of the fittest'. This was actually coined by the philosopher and political thinker Herbert Spencer, although Darwin did borrow it for later editions of the *Origin*.
8. Perhaps the best known of these calculations was by James Ussher (1581–1656), the Archbishop of Armagh, who, in 1650, published his work showing that the Creation occurred on 23 October 4004 BC.
9. Darwin is alluding to the lines from A Midsummer Night's Dream spoken by Oberon towards the end of Act Two Scene 1:

    I know a bank were the wild thyme blows
    Where oxlips and the nodding violet grows
    Quite over-canopied with luscious wood-bine,
    With sweet musk-roses, and with eglantine.

10. Gorillas had only been described by western science a few years before the *Origin* was published and the first complete specimen was seen in London in 1861. The discovery of the gorilla caused something of a sensation in Victorian Britain, one of the reasons why satirical magazines such as *Punch* carried so many cartoons depicting Darwin with a gorilla's body or ran articles suggesting he had a gorilla for a grandfather.
11. Mount House is now known as Darwin House. It is not a

museum, but is used by the local council to house the Shrewsbury Valuation Office, although it can be visited during normal office opening hours by anyone interested in seeing Darwin's birthplace.

12. FitzRoy had originally brought four Fuegians to England, although the methods he used to do this would better be called kidnapping. The Fuegians were given the English names Jemmy Button, York Minster, Fuegia Bucket and Boat Memory and they became minor celebrities in Britain before three of them returned to Tierra del Fuego, one, Boat Memory, having died of smallpox in England.
13. The publication of transcriptions of the notebooks, no easy job given Darwin's cryptic handwriting, by Barrett et al (1987) was something of a landmark in Darwin studies. The notebooks, along with all sorts of other manuscripts and documents were found in a cupboard under the stairs in Down House after Emma Darwin died in 1896 and they offer an extraordinary glimpse into the development of the theory of natural selection as well as showing how wide-ranging his reading and thoughts were during the period the notebooks cover, roughly from late 1836 to 1844.
14. The Whig party came to power in the 1830s and based their policies of providing for the poor on the writings of Malthus, resulting in the Poor Law Amendment Act of 1834 and the establishment of the deeply unpopular work-houses. Harriet Martineau was a keen advocate of these policies and, as a close friend of his brother Erasmus, may have been the person who encouraged Darwin to read Malthus at this time.
15. Before his death in 2002, Stephen Jay Gould wrote more than 300 essays, many of them published in *Natural History* magazine in America and subsequently collected in a series

of books. The essay referred to here is 'Worm for a Century, and for All Seasons', first published in April 1982, the month of the centenary of Darwin's death, and can be found in Gould (2006).

16. This quote is from an essay called 'The Most Unkindest Cut of All', which is also reprinted in Gould (2006).
17. As well as Stephen Jay Gould, notable critics of sociobiology and evolutionary psychology have included Steven Rose and Richard Lewontin. Well known names from the other side of the argument include Richard Dawkins, Steven Pinker and Daniel Dennett.
18. Gregor Mendel, it should be noted, did not know the physical basis of heredity, but observed that organisms inherit traits from their parents in a discrete manner and did not blend the traits of each parent which was the accepted explanation at the time.
19. A straightforward account of the breadth of evolutionary research is given in Zimmer (2003), while, for those with plenty of time on their hands, the 1,400 pages of *The Structure of Evolutionary Theory* by Stephen Jay Gould give a full account of evolutionary thought and his own theory of the punctuated equilibrium. Although now more than 30 years old, *The Selfish Gene* by Richard Dawkins remains as good an account as any of evolution at the level of the gene and also provides some balance to Stephen Jay Gould's work, which stresses the evolution of the phenotype rather than the genotype. More recent advances include evolutionary developmental biology, or evo devo for short, and epigenetic inheritance systems. While a bit beyond the scope of this book, good accounts are given in *Endless Forms Most Beautiful* by Sean Carroll and *Evolution in Four Dimensions* by Eva Jablonka and Marion J. Lamb respectively.

20. The study by the Grants, related in the Pulitzer Prize winning book *The Beak of the Finch* by Jonathan Weiner, built on another classic study on Darwin's finches conducted by David Lack in the 1940s.
21. Other notable titles include *Breaking the Spell: Religion as a Natural Phenomenon* by Daniel Dennett, *God is Not Great: How Religion Poisons Everything* by Christopher Hitchens and *The End of Faith: Religion, Terror and the Future of Reason* by Sam Harris.
22. The stage play *Inherit the Wind* is a fictionalised account of the Scopes Trial, which was subsequently made into a film of the same name starring Spencer Tracey.
23. The full 139 pages of the decision by Judge John E. Jones in the Dover Trial is available online at www.pamd.uscourts.gov The judge, a republican appointee of President George W. Bush and regular church goer, was accused by the intelligent design movement of being an activist bent on preventing the spread of a legitimate scientific theory, an accusation he strongly rebutted, going on to criticise the ID movement in a way that would have made Richard Dawkins proud.

Books by Charles Darwin

1838–1843: *The Zoology of the Voyage of H.M.S. Beagle*. 5 vols. *edited by Charles Darwin*

1842: *The Structure and Distribution of Coral Reefs*

1844: *Geological Observations on the Volcanic Islands Visited During the Voyage of H.M.S. Beagle*

1845: *Journal of Researches into the Natural History and Geology of the Countries Visited during the Voyage of H.M.S. Beagle Round the World*. This was originally published as the third volume of *The Narrative of the Voyages of H.M. Ships Adventure and Beagle* and an edited version is now known as *The Voyage of the Beagle*.

1846: *Geological Observations on South America*

1851–1854: *Living Cirripedia (Barnacles) 2 vols.*

1851–1854: *Fossil Cirripedia 2 vols.*

1859: *On the Origin of Species by Means of Natural Selection, or the Preservation of Favoured Races in the Struggle for Life*

1862: *On the Various Contrivances by which British and Foreign Orchids are Fertilised by Insects*

1865: *On the Movements and Habits of Climbing Plants*

1868: *The Variation of Animals and Plants under Domestication 2 vols.*

1871: *The Descent of Man, and Selection in Relation to Sex 2 vols.*

1872: *The Expression of the Emotions in Man and Animals*

1875: *Insectivorous Plants*

1876: *The Effects of Cross and Self Fertilisation in the Vegetable Kingdom*

1877: *The Different Forms of Flowers on Plants of the Same Species*

1880: *The Power of Movement in Plants*

1881: *The Formation of Vegetable Mould, through the Action of Worms*

Bibliography

Armstrong, Patrick H., *All Things Darwin: An Encyclopedia of Darwin's World,* 2 vols, Connecticut: Greenwood Press, 2007

Bowler, Peter J., *Evolution: The History of an Idea*, Revised edition, California: University of California Press, 1989

Browne, Janet, *Charles Darwin: Voyaging*, London: Jonathan Cape, 1995

Browne, Janet, *Charles Darwin: The Power of Place*, London: Jonathan Cape, 2002

Browne, Janet, *Darwin's Origin of Species: A Biography*, London: Atlantic Books, 2006

Darwin, Charles, *Autobiographies*, Harmondsworth: Penguin, 2002

Darwin, Charles, *Charles Darwin's Notebooks, 1836 – 1844*, ed. Barrett, Paul H., Gautrey, Peter J., Herbert, Sandra, Kohn, David and Smith, Sydney, Cambridge: CUP, 1987

Darwin, Charles, *The Correspondence of Charles Darwin Volume 7 1858 – 1859*, ed, Buckhardt, Frederick and Smith, Sydney, Cambridge: CUP 1991

Darwin, Charles, *The Descent of Man*, Harmondsworth: Penguin, 2004

Darwin, Charles, *The Origin of Species*, Harmondsworth: Penguin, 1968

Darwin, Charles, *Voyage of the Beagle*, Harmondsworth: Penguin, 1989

Dawkins, Richard, *The God Delusion*, London: Bantam Press, 2006

Dawkins, Richard, *The Selfish Gene*, Oxford: OUP, 1976

Desmond, Adrian and Moore, James, *Darwin*, London: Michael Joseph, 1991

Dennett, Daniel C., *Darwin's Dangerous Idea: Evolution and the Meanings of Life*, Harmondsworth: Allen Lane, 1995

Gould, Stephen Jay, *The Richness of Life: The Essential Stephen Jay Gould*, London: Jonathan Cape, 2006

Hodge, Jonathan and Radick, Gregory (eds.), *The Cambridge Companion to Darwin*, Cambridge: CUP, 2003

Mayr, Ernst, *One Long Argument: Charles Darwin and the Genesis of Modern Evolutionary Thought*, Harmondsworth: Allen Lane, 1992

Richards, Robert J., *Darwin and the Emergence of Evolutionary Theories of Mind and Behaviour*, Chicago: University of Chicago Press, 1987

Ridley, Mark (ed.), *Evolution*, 2nd Edition, Oxford: OUP, 2004

Zimmer, Carl, *Evolution: The Triumph of an Idea: From Darwin to DNA*, London: Heinemann, 2002

Index

Agassiz, Louis, 43
Amazon rainforest, 18, 146
American Civil Liberties Union, 138
Andes, 90–92, 146
Anglican Church, 31, 45, 47, 58, 67, 101
antibiotic resistance, 133
Argentina, 84, 86, 88
argument from design, 10, 139, 140, 142–143
Arkansas, 139
Ascension Island, 97–98
atheism, 136, 145
Atlantic ocean, 75, 90, 97
Australia, 97
autobiographical writings, 43, 63, 70, 102, 104
Azores, 97

Bacon, Francis, 51, 102–103
Bahia, 77, 83, 85, 97
barnacles, 13, 105, 146
Bates, Henry Walter, 18
Beagle, HMS, 10, 13, 16, 18, 29, 34, 40, 42, 44, 50, 52, 64, 69, 71–74, 76–77, 79–80, 82–84, 88, 90–94, 97–103, 111–112, 145, 153
Beaufort, Francis, 75–76
beetles, 65, 68, 83
Bible, the, 40, 47–48, 133
biological determinism, 122
biology, 28, 48, 103, 125–127, 129, 133–134, 136, 150
Boer War, 120
botany, 67, 108, 111, 116
Brazil, 77, 83–84, 89, 97
Britain, 13, 19, 37, 39–40, 42, 44, 57–58, 88, 98, 108, 114–115, 117, 120, 137, 143, 148–149
British Association for the Advancement of Science, 46
Buenos Aires, 84–85, 88, 92
Buffon, Comte de, 8, 57
Butler Act, 138

Cambridge University, 24, 31, 41, 62, 65, 67, 69–70, 78–79, 82–83, 98
Canary Islands, 69
Cape Verde Islands, 79–80, 103
Castlereagh, Viscount, 75, 100
catastrophism, 81
Chambers, Robert, 38
Chile, 90, 92, 96
Christianity, 87–88, 131–132, 144
chromosomes, 126
chronometers, 71, 74
climbing plants, 109
Cocos (Keeling) Islands, 97
co-evolution, 108
Copernicus, 36, 131
coral, 13, 97, 112–113, 146
creationism, 10, 137–142
Crick, Francis, 127
Cuvier, George, 81

Darwin, Annie, 16, 24, 144
Darwin, Emma, 15 - 17, 64, 101, 113, 149

Darwin, Erasmus (brother), 56, 61, 149
Darwin, Erasmus (grandfather), 8, 57–59
Darwin, Henrietta, 17, 22
Darwin, Leonard, 120
Darwin, Robert, 56, 58–59, 61–62, 64, 72–73
Darwin, Susannah, 56, 60
Darwinian evolution, 7–8, 10, 44, 119, 123–126, 133, 145
Dawkins, Richard, 11, 44, 123–124, 134–136, 150
Descent of Man, The, 23, 39, 109–110, 116, 119, 154
DNA, 124, 127, 133
domestic animals, 32, 80
Dover Trial, 140
Down House, 14–15, 23–24, 26, 28, 32, 44, 107, 147, 149, 156

Earth, the, 9, 34–35, 81, 95, 131
earthquakes, 82
earthworms, 111–113, 146
Edinburgh University, 57, 61–62
Eldredge, Niles, 128
essay of 1844, 21, 105
Essay on the Principle of Population, An, 104
eugenics, 48, 117–121
evolution, 8–11, 18, 20, 22, 26, 28, 30–31, 34, 36, 38–39, 41, 43–44, 46–49, 57, 63, 107–112, 115–117, 125, 128–129, 131–145, 147–148, 150, 156
evolutionary psychology, 122–123, 150
Expression of Emotions in Man and Animals, The, 110
extinction, 89

Falkland Islands, 88 - 90, 95
finches, Darwin's, 32, 96, 103, 128, 132, 151
FitzRoy, Robert, 40, 47, 71–81, 83–85, 87–88, 90–92, 97, 100, 103, 149
Formation of Vegetable Mould, through the Action of Worms, The, 111, 154
fossil record, 128, 132
fossils, 42, 84, 99, 128
Fox, William Darwin, 24, 65, 67
French Revolution, 57
fundamentalism, 10, 48, 133

Galapagos islands, 32, 78, 80, 91, 93–94, 96, 103, 128, 146
Galileo, 131
Galton, Francis, 48, 116–119, 122
genes, 36, 122–124, 127
Genesis, 40, 48, 133
genetics, 36, 126, 127
genocide, 121
genotype, 126, 150
Geological Society of London, 99
geology, 41, 51, 62, 68–70, 80–82, 87, 89, 92, 94–95, 100, 116
Germany, 120
God, 10, 31, 49, 66, 81, 131–132, 135–136, 139–140, 145, 151
God Delusion, The, 136
gorillas, 45, 148
Gould, John, 99, 103
Gould, Stephen Jay, 111, 121, 128, 134–135, 149, 150
gradualism, 34, 81–82, 112, 128
Grant, Peter and Rosemary, 132, 151
Grant, Robert, 63
Gray, Asa, 21, 43, 143
Great Exhibition of 1851, 38

Henslow, John, 67–71, 79, 99
heredity, 150
Herschel, John, 68, 97–98
Holocaust, 121
Homo erectus, 124
Homo habilis, 124
Hooker, Joseph, 13, 20–23, 43–44, 47, 108, 113
human evolution, 39, 116

INDEX

human genome, 127
Humboldt, Alexander von, 69, 79, 98
Hume, David, 51–53, 142–143
Huxley, Julian, 126
Huxley, Thomas, 8, 49, 126, 144
hydrotherapy, 24, 26

immutability, 31
Indian Ocean, 97
industrial revolution, 37, 58
inheritance of acquired characteristics, 50, 142
insectivorous plants, 111
intelligent design, 48, 139–141
irreducible complexity, 139–140, 142
islands, 34, 80, 87, 89, 93–97, 103, 146

John Paul II, Pope, 132
Journal of Researches, 13, 25, 69, 153

kelp, 89–90, 146
Kew Gardens, 20, 108
King, Philip Gidley, 82
Kipling, Rudyard, 115

laissez-faire, 47
Lamarck, Jean-Baptiste, 8, 49, 57, 63, 142
Lamarckism, 125
Linnean Society, 21, 105
London, 14, 27, 37, 38, 42–44, 99, 101, 107, 120, 147–148, 156
Louisiana, 10, 139, 141
Lyell, Charles, 13, 18–21, 25, 34–35, 40–44, 81–82, 92, 98–99, 101–102, 113, 143

Maer, 60, 64, 70, 72, 101
Magellan Strait, 90
Malay Archipelago, 18
Malthus, Thomas, 33, 53, 66, 104, 117, 149
Martineau, Harriet, 149
Marx, Karl, 7
memes, 123–124
Mendel, Gregor, 36, 126, 150
Mental Deficiency Act 1913, 120
Milton, John, 29, 148
modern synthesis, 36, 126–127, 132
Montevideo, 85
Murray, John, 25–27, 107, 110, 148
mutation, 126, 129

natural selection, 8, 10, 21, 23, 28–34, 36, 39, 49–50, 52, 66, 81, 90, 93–94, 105, 108–109, 114, 117, 119, 121, 125, 143, 149
nature versus nurture, 122
Nazis, 121
Neanderthal man, 124
New Zealand, 97
Newton, Isaac, 36, 113, 131
NOMA, 135
North Wales, 69

orchids, 107–109, 146
Origin of Species, The, 7, 9, 17, 23, 25–26, 28, 36, 43, 49, 63, 66, 87, 93, 106–107, 128, 131, 145–147, 153, 156
Owen, Richard, 42–43, 45, 64, 99

Pacific ocean, 90, 92–93, 95, 97
Paley, William, 65–66, 142–143
pampas, 85–86, 146
pangenesis, 109
peacocks, 110
peas, 36, 126
phenotype, 126–127, 150
philosophy, 51–53, 103, 136
pigeons, 13, 32
Pius XII, Pope, 132, 135
political theory, 51–52, 123, 143
population theory, 66
punctuated equilibrium, 128–129, 150

race, 110, 116
racism, 125
radicalism, 30, 57–58, 62, 143
religion, 10, 40, 64, 123, 131, 135–137, 144
rhea, 86–87
Rio de Janeiro, 84
Roman Catholic Church, 132
Royal Navy, 37, 72, 74–75, 84–85

Science Education Act, 141
Scopes Monkey Trial, 10, 138, 151
Scopes, John, 138
Second World War, 121
secular humanism, 136
Sedgwick, Adam, 41, 67, 69–70, 79, 98–99
Selfish Gene, The, 123, 150
sexual selection, 110, 116
Shakespeare, William, 29, 35
Shrewsbury, 56–62, 65, 69–70, 73, 90, 98, 149
Shrewsbury School, 61–62, 65, 90
slavery, 46, 76–77
Smith, Adam, 52, 114
social Darwinism, 115, 125
sociobiology, 122–123, 150
speciation, 96
Spencer, Herbert, 47, 114–116, 148, 151
St. Jago, 80, 82–83, 103
sterilisation, 120–121
Stokes, John Lort, 82
struggle for life, 90, 93
survival of the fittest, 47, 115, 143, 148

tabula rasa, 122
Tahiti, 97
teleological argument, 48, 142
Tenerife, 69–70, 79
Tennessee, 138–139
Tierra del Fuego, 74–75, 87–90, 146, 149
tiktaalik fish, 133
tortoise, Galapagos, 96
Transmutation Notebooks, 102
transmutation of species, 102
tree of life, 9, 110

Uniformitarianism, 81
Unitarian church, 58, 60
United States of America, 140
University College London, 120
US Constitution, 137, 140–141
US Supreme Court, 138–140
USA, 71

Valparaiso, 90–91
variation, 8, 29–30, 32–33, 36, 102, 125, 129
Vestiges of the Natural History of Creation, 38
volcanoes, 82, 91–92
Voyage of the Beagle, 13, 69, 153

Wallace, Alfed Russel, 17–23, 105–106, 147
warrah, 89
watchmaker analogy, 142
Watson, James, 127
Wedgwood, Josiah (grandfather), 16, 37, 56, 60
Wedgwood, Josiah (uncle), 60, 73
Westminster Abbey, 113
Westminster Review, 45
Wickham, John Clements, 91
Wilberforce, Samuel, 46
Wilson, Edward O, 122
women's suffrage, 116

Zoology of the Voyage of HMS Beagle, 99, 153